*Organizing Science for Technology Transfer
in Economic Development*

ORGANIZING SCIENCE

FOR TECHNOLOGY

TRANSFER

IN ECONOMIC

DEVELOPMENT

ROBERT A. SOLO

MICHIGAN STATE UNIVERSITY PRESS

1975

★
★
★
★
★

TO
Rosie la douce

Contents

Acknowledgments

I am obliged to the International Programs at Michigan State University, particularly to Ralph Smuckler, Richard Niehof, and Dole Anderson, through whom I obtained the MUCIA Ford Foundation Grant that permitted my research in Europe during the fall of 1971, and to Professor Fritz Machlup under whose auspices I spent the year 1965–66 at Princeton, where I completed a first draft of this book.

The study itself is based primarily on visits to and interviews at government agencies and research and development centers. Those contacted were nearly always cooperative. Particularly I remember with gratitude the a number of people.

In Great Britain; I thank Graham Jones for his guidance; Ray Beverton, P. T. Haskell, F. C. Bawden, Iean Maddox, Professors Kearns and Wilson for their insights into the specifics and politics of the matter; Sir Thomas Scrivnor and Anthony Tasker for their great courtesy; Ray Millard and Dr. Tingle for putting me on the road; G. R. Ames, Philip Spencley, and Stanley Hiscocks, who led me through the ins and outs of the T.P.I.; Directors Astbury, Allen, Adams, Cutting, Hicks, Stout, Lawrie, and Sopwith for enlightening me concerning their respective research associations and laboratories.

In Germany, I am especially grateful for the help and kindness of Professor Hans Havemann of the Research Institute for International Techno-Economic Cooperation at the University of Aachen, and Hans Harborth, also of that Institute. Herren Stecher and Stangen in the ministries and Mensching at GAWI were most helpful. I learned much about German fisheries from Professors Von Brandt and Herr Seemann; about German forestry from Professor Kubiena, Herr Jorden, and Herr Mammen; about geology in Germany from Professor Richter Benberg; and about the science infrastructure from Professor U. Stille.

In the Netherlands, I owe a great debt for the devoted help of my friend J. C. Gerritsen of the TNO. I enjoyed a most useful discussion with Professor Van Dam on Dutch science policy. In studying the international corporation as a change agent, I gained useful insights from Mr. Veraart of Verenidge Motorfabrietien, Mr. Boer of Royal Dutch Textiles, and Mr.

Vander Brink of Philips. At the International Agricultural Center, Mr. Haak and Mr. P. A. Blijdorp, at NEDECO, Mr. Frijlink, and at the Research Institute for Management Sciences, Mr. Van Veen were all most helpful.

In France, I thank Bernard Cazes, who is truly a friend of the American scholar; M. Berthelot of INSE, then secretary of the "Comité des Sages," Messeurs Aubert, Delais, Camus, Maigien, Rodier, Sengalin, Fournier of ORSTOM; Carrier de Belgerac of the IRHO; Lesche of the BDPA; Boor of SATEC and IRAT; Bourriers of BCEOM; Pagot of the Institute of Tropical Veterinary Medicine; St. Hilaire of IRCT; Coste of IFCC, Beaumont and Mesenin of BRGM; and Mme Landowski of the Ceramics Institute.

Organizing Science for Technology Transfer
in Economic Development

Introduction

AN EXPLANATION OF HOW THIS BOOK CAME TO BE WRITTEN WILL HELP the reader understand what it is about, its message, its limitations. In 1962 my article "Gearing Space-Military R & D into Economic Growth," in the *Harvard Business Review,* impressed a number of people in government and business circles, and gained for me a small flash of fame, a momentary light that attracted and evidently blinded an aggressive official on the science policy side of the Organization for Economic Cooperation and Development (OECD), who cabled me urgent offers of work for that organization in Paris. I agreed to go for a year and, in due course, I arrived at the Organization's headquarters. It was spring. Paris was lovely. The OECD, alas, was something else again.

This was my first and only experience in an international agency, and yet I hardly think that that one can be entirely like the others. The OECD is something between the United Nations and the European Economic Community. Except that the UN and the EEC have real or potential functions: the UN to settle conflicts between power blocs, the EEC to organize and preserve a very complicated set of economic relationships. The OECD has no function. It is a diplomat's vanity, a conceit of the U.S. State Department. But to say that it is without purpose is not quite to say that it is without value. It organizes numerous conferences. At these conferences medium and high officials, even ministers, from the countries of Western Europe and the United States, and more recently from Japan, meet in the grand ballrooms of a former Rothschild mansion to exchange grand banalities on such grand subjects as Education, Science, Economic Development, Environmental Improvement, or whatever happens to be the high intellec-

1

tual fashion of the instant. Useless? Not necessarily. Some officials who attend some of these conferences may learn something or other in the ballrooms or at the bar. Predictably, they arrive at this single grand conclusion: more studies are needed. In the making of these studies a large bureacracy, loaded with experts and consultants, has developed at the OECD. Useless? Not necessarily. Some studies are informative. Some experts are helpful. And some officials, bureaucrats, consultants, experts—clothed with organizational status and using the established communications outlets— exert an influence, benign or malignant, on other individuals and organizations. It could hardly be otherwise.

Who pays? The United States is Big Daddy, presumably through the corporate person of the State Department and the large foundations. Who controls? The bureacracy, at least that part of it with which I was connected, was run by a tight, tough British clique.

Given a large bureacracy, with no unifying purpose or explicit function, composed of individuals desperately anxious to stay on the payroll (and who can blame them when the pay is high and a life in the City of Light is at stake), it is easy to understand why vast energy should be devoted to the low, backstabbing politics of personal survival. Less easy to understand is the secrecy and spying, the backroom censureship, the searching through private files, the reading of private mail. Paranoia perhaps, but this was the only time in my career when I suspected my associates and superiors of being undercover agents for the CIA or its British equivalent.

Do these words suggest that I was persecuted, harassed, and unhappy? On the contrary, it was a most delightful fifteen months. I had nothing to do. I was the highest paid member of the division, yet I was never given an assignment. Indeed, when I was finally given a desk, it was in the passageway to the storeroom. No assignment, no office, high pay, and in Paris—a more perfect situation could hardly be imagined. Why did they bring me? I am not sure. Perhaps what happened was that a rival of the man who hired me threatened to steal the spotlight by organizing the first conference of the ministers of science. To outshine this rival, I was to produce a second larger, longer, more glamorous conference. By the time I got to Paris, the rival was on his way out, and the plan for the second conference had been shelved.

Alas, I am haunted by a sense of duty. Since I was being paid, I expected to work. I even yearned to be useful. Besides, I wanted to find out for myself more about what was going on in Europe. So, in spite of my bosses, I made work, traveled, researched and wrote, and tried to stir up some action.

My division was concerned with science policy. I was brought in as an expert on science. One of the major interests of the OECD is in under-developed countries. I am also interested in economic development. The linking of science policy and economic development, therefore, came easily. Just a few years earlier I had conceived and started a program in Puerto Rico to organize science resources there for economic development. Hence, toward the end of my tenure at the OECD, I started, quite alone and on my own, a study of how France, Germany, the Netherlands, and Great Britain organized their science resources as part of their programs to pro-mote economic development in underdeveloped countries of the world.

Since even sophisticated readers can be confused by them, an explanation of some of the terms used in this study is necessary. The contributions of a rich, technically advanced country intended to raise the level of productiv-ity in a poor, technically backward one will be referred to as AID. The giver of AID is a *donor*. The poor country receiving AID is a *recipient*. AID recipients have been variously designated as low-productivity countries, underdeveloped countries, and, currently, as "less-developed countries," or *LDCs*. Donors contribute cash grants or loans made allegedly at a low rate of interest and on easy terms, i.e., *capital* or *financial* AID; or their contri-bution may take more concrete and specific forms, for example, in the sending of experts, machines, equipment, and otherwise undertaking to upgrade LDC technology. By *technology* is meant the organized capability for some purposeful activity. AID in upgrading (or installing new) tech-nology is called *technical assistance*. The interest of this book is in one aspect of technical assistance, namely, the use by donors of their scientists, research centers, and other science-based organizations, i.e., the use of their *science resources* for the purpose of technical assistance. Hence, the book has to do not only with AID but also with government policies related to the use of science resources—hence, with *science policy*.

In the final study made during the last months of my tenure with OECD, under pressure and in haste, I visited key government agencies and the research establishments in France, Germany, the Netherlands, and Great Britain, rushed out a draft report and turned it over to my bosses. Their response was curious. At first they welcomed and praised it. Then they turned suspicious, sour, and fearful, though no question was ever raised with me as to the accuracy of my observations or the value of my conclusions. The division chief ordered that the final chapter, "Conclusions and Recommendations," be torn off all the several hundred mimeographed copies and summarily destroyed. No remnant was allow to remain! The hand-written manuscript from which the report had been typed and mimeographed was sequestered and I was denied access to it. It, too, presumably was destroyed, and subsequently I have never seen the offending conclusions and recommendations, nor have I attempted to reconstruct them. What remained of the report, the OECD has kept hidden. Only a select few have been permitted to see it. Even today, I am told, it is kept as as a secret document, treated as something dangerous, subversive, or God knows what! And years after I had left the OECD, my former bosses applied crude pressure to keep me from publishing on the subject. I have never understood what caused such excitement, but the their reactions gave me reason to suppose that I must have said something worth saying.

I left the OECD in the early fall of 1964. There followed a jumble of events, all keeping me from finishing what I had started in the study of science-based AID. By the time I got back to that task, my data was seven years old. So, with the benefit of a MUCIA grant from Michigan State University, I returned to Europe during the fall of 1971 to resurvey the science-based AID programs in the countries I had studied, to update my information, and to gain some historical perspective on drift and change in science and in AID policies and programs since 1964.

The limitations and biases implicit in this study should be made explicit. Its evaluation of science-based aid programs is one-sided, necessarily so, since I have studied only the organization of these programs in donor countries, but not their implementation in LDC recipients.

Nor was equal time devoted to nor could I approach with equal authority the analysis of science-based AID in each of the countries. I had not the opportunity to study the organization of science-based AID in Germany in the same depth as I had, say, in France. Moreover, many important donor nations—large ones like the United States, the Soviet Union, China, and Japan, and smaller ones like Canada, Belgium, Sweden—are not included. The four surveyed, however, are the most important of the West European donors. It would, in fact, be difficult to pose the same question concerning AID policies in the Soviet Union or the United States that I do about West European donors. Consider, for example, the United States.

"How does government mobilize and organize science resources in AID programs?" is a meaningful question when it is asked concerning the four West European countries. Not so for the United States. The Agency for International Development in the United States, through all its variations and permutations, from its glorious and hopeful beginnings to its dismal quasi-demise, has never deliberately mobilized, organized, or deployed science resources as part of technical assistance. It has been innocent of any such intention. It has been without any such conception. Nor has any other agency of American government undertaken or acquired the competence to mobilize science resources or to organize research activities for such purposes. Where such initiatives have been taken in the United States, it has been by the great foundations, sometimes by academics operating autonomously in the university. As for the political authority, there has been no science policy, no mobilization of science resources as an aspect of AID.

In American experience (and that is inevitably a reference base for this study), important changes came about in attitudes toward AID in the period between 1963, when I began this study, and 1972, when I completed it. The year 1963 was at the high tide of enthusiasm and expectation. Jack Kennedy was president. Youth was in the saddle. The Peace Corps was created. The Alliance for Progress was celebrated. Its council of wise men, drawn from among the nations was to review and approve LDC development plans, which a beneficent American government would then support with guidance and gold. Teodoro Moscoso was brought from Puerto Rico to put the *Allianza* on the road. Americans received the brand-new set of Keys to the Kingdom with euphoria.

However, by 1971 the Keys had been lost, misplaced somewhere: they had opened no Golden Gates. The high hope was gone. The wise men had proven to be not very wise. The enthusiasm was lost. AID had dwindled; its future was in doubt.

What was the anatomy of disillusion? It was compounded of the failures and waste of donor programs; of the ugly politics, arrogance, fanaticism, brutal oppression, and terrible cruelty displayed by the new and poor nations themselves: by Nigeria, Brazil, Algeria, Mexico, India, Pakistan,

Egypt, Syria, Greece, Turkey, Argentina, and the rest—so that the poor guys no longer seemed like the good guys. On the contrary, we could give thanks that they were not rich and powerful enough to have their way with the world. And the American good will had been perverted, its idealism degraded by the hard-nosed, cold warriors of the State Department, with their obsessive anticommunism. Under the cover of AID, they had used political graft and billy clubs, CIA ruffians and military advisers, machine guns, jet aircraft, tanks, assassination and wars—to support the most corrupt and repressive dictatorships, to increase the wealth of the fabulously wealthy, to oppose any significant social change, to crush all revolutionary spirit, ending their adventure in the morass of Vietnam. During those nine years America's position in the world had indeed changed gravely. Her image and self-image was tarnished, more than tarnished—befouled. Her claim to moral leadership was gone. Even pretension of superior know-how and of a higher economic wisdom was vanishing, for by 1971 the pace of American technical advance too was lagging. The United States was being overtaken in industrial productivity. The dollar had lost its place among the currencies. During these years there was but one outstanding success in technical assistance—the so-called Green Revolution.

In 1922 the Japanese introduced into Taiwan ponlai rice varieties, a very short-stemmed plant with a minimal foliage that, when used with a heavy input of fertilizer and controlled water supplies, could yield an extraordinary crop. Although by 1940 the Japanese had developed disease-resistant strains and had organized the system for its cultivation, they never really benefited from their innovation. During the war a critical shortage of chemical fertilizer prevented its exploitation, and after the war possession passed to the Chinese Nationalists. In postwar decades Taiwanese rice culture produced quite unprecedented per acreage yields.[1]

During the postwar decades the Rockefeller Foundation established in Mexico a research unit CIMMYT (International Maize and Wheat Improvement Center) to promote agricultural advance in that region. Following the Japanese prototype, the Center developed a (Mexipac) wheat variety, which turned out to be well adapted to climatic and soil conditions on the Indian subcontinent also. There its use was promoted in cooperation with the Ford Foundation. The subsequent increase in the acreage cultivating this new wheat variety was truly phenomenal—the 200 acres planted in 1964–65 became 34,000,000 acres in 1968–69.[2] By the end of the decade, India and Pakistan were on their way to becoming self-sufficient in wheat production, or even exporters of the grain. The significance of this amazing change was dramatized by India's demonstrated ability to feed the millions of Bangladesh refugees and yet sustain her own people.

CIMMYT continues to build up "gene pools covering such factors as disease, insect and drought resistance, protein quantity and quality, and insensitivity to day length, providing basic raw materials from which national breeders around the world can develop superior varieties for their specific conditions."[3] In 1962 the IRRA (International Rice Institute) was established under the auspices of the Ford and Rockefeller foundations. CIAT (International Center for Tropical Agriculture) in Colombia and the IITA (International Institute of Tropical Agriculture) in Nigeria are in process of formation, receiving support from the recipient governments as well as from the two foundations.

The Green Revolution has some interesting implications for economic development planning.

1. It is foresquare *science-based,* produced at a low cost through the deliberated and sustained research and development efforts of a small group of research workers; yet it dwarfs into insignificance the benefits of the billions of donor dollars spent as financial assistance and capital expenditures, or even technical assistance in building steel mills and automobile plants and roads and highways and the rest, that too often has produced only a crop of millionaires in the LDCs who pay their workers starvation wages and sell a heavily protected output at very high prices. Science-based technical assistance has been entirely neglected by the United States government in its AID programs.

2. As a scientific-technological achievement, the development by the CIMMYT of the new wheat variety was merely routine. The Japanese had developed and proven out both the prototype plant and the system for its cultivation. The value of both had been established by Taiwanese experience. The Rockefeller researchers had only to follow in the Japanese path. It was a matter of imitation and adaptation.

   Hence the question: why so slow? From 1922 until 1965 is nearly half a century. In the face of this fact (and so many others like it), what are we to make of the lightning speed with which scientific information is alleged to move by the journal route from mind to mind throughout the world?

Why had not this essentially routine research task, of such real and urgent importance, been achieved by others? Why had it not (or anything like it) been achieved by those powerful research institutes in France, oriented to economic development, or by research in Britain or Germany or the Netherlands? Why had it not been done (indeed what, if anything, has been done?) by the research institutes established in the LDCs themselves? India, whose fate was at stake, had a plethora of scientists. Its atomic research establishment, employing thousands, is among the most lavishly equipped in the world. If the research of that establishment has in any way contributed to economic development or human betterment, the fact has been well hidden. Yet the Indians never undertook the simple research task required for the Green Revolution, which saved them body and soul. Why not?

3. The Green Revolution is an agricultural revolution. The primary aim of development planners in donor and in recipient countries alike had been to industrialize. Their formulas were sound money, high saving ratios and capital investments, and birth control. The high-cost factories the planners established in these rural lands mattered little to the peasant mass, treading at starvation's edge.

4. Another side of the story is yet to be told. The new seed varieties required a quite different sort of cultivation, which had to be learned. The new cultivation required a carefully controlled supply of water. Hence, tube-wells, pumps, or water reservoirs and an irrigation system had to be available. The cultivation of new varieties required insecticides and the massive use of fertilizer. What is extraordinary is that the necessary infrastructure was in place and that those Asian peoples were ready to receive and were able rapidly to implement the new potential.

Any major technological transformation creates problems that are the inverse of the possibilities for betterment. When more grain is produced, it must be transported and distributed. If the transportation and distribution system is not readied, there will be crisis and despair. When large-scale agriculture becomes profitable, land holdings will be consolidated and the tenant will be pushed off the land. The labor force, released from the land, provides the potential basis for an industrial work force and, hence, for the industrial enrichment of society. But if there is not a corresponding capacity to absorb the worker into industry, there will be crisis and despair. Increased grain and other food outputs will reduce their price, to the benefit of consumers. But an uncontrolled price plummeting downward can bankrupt the farmers and disarrange agricultural organization and disrupt its progress, also causing crisis and despair.

The Green Revolution affected the outlook of development planners throughout the world. In 1971 all the questions that it raised and all of its implications for development planning were still hanging in the air.

NOTES AND REFERENCES

1. Vernon Ruttan "Planning Technological Advance in Agriculture: The Case of Rice Production in Taiwan, Thailand and the Philippines," in Robert A. Solo and Everett M. Rogers (eds.) *Including Technological Change for Economic Growth and Development* (East Lansing: Michigan State University Press, 1972).

2. Lester C. Brown, *Seeds of Change* (New York: Praeger, 1970). Published for the Overseas Developmental Corporation. Mr. Brown's book has received attention because its publication coincided with the news of the Green Revolution. Curiously, Mr. Brown wrote the book without an inkling of the source and history of the innovation of which he makes so much. If his book is empty of analytic value, it is full of institutional propoganda and is pernicious in the illusions it seeks to preserve.

Thus, in relation to "The Role of the United States Government," we are told (pp. 66–67), "A lot of foreigners and a lot of foreign aid went into the making of the agricultural revolution, but American institutions provided by far the largest contribution, whether measured in terms of men or money. And American foreign aid was centrally involved. The role of the American government in helping to transfer the technologies that made the agricultural breakthrough possible can be compared to the role that the American government played in the Marshall Plan." Mr. Brown then spells out the "role that the Agency for International Development played in helping to transfer the technologies that made the agricultural revolution possible." There follows the familiar jumble of AID items to be found in any agency press release, from which no specific impact upon nor value for development can be deduced. What is interesting about this alleged garnering of the facts and spelling out of the contribution of USAID is what is unsaid. For Mr. Brown's book makes not the slightest claim that USAID or any other agency of the American government supported or was even aware of the R and D (Research and Development) that produced the Green Revolution. That scientific-technological breakthrough was a Japanese achievement, adapted and disseminated by a private foundation. No claim is made, for no claim could be made, that USAID (or the U.S. Department of Agriculture) organized and supported any R and D program whatsoever to promote agricultural or industrial development among the LDCs. At the level of development-oriented science and the systematic adaptation of technology, USAID has been a nullity.

3. Graham Jones *The Role of Science and Technology in Developing Countries,* (New York: Oxford University Press, 1971), p. 91.

Chapter 2

Agents of Change

AS BETWEEN SOCIETIES, THE DIFFERENCES IN PRODUCTIVITY, AND hence in income, are not marginal, nor do they shade gradually one into the other. Rather, there are distinct sets with an enormous gap between them. One set, the LDCs, constituting two-thirds of the world's population, had an average per capita annual income of $135 in 1969. The other set, with one-third of the world's population, had an annual per capita income of $1,800.[1] Closing this gap is what AID is about. To close it requires a fundamental change in the way that the LDCs organize their production processes. Hence, AID can be understood as a way of changing the technology of low-productivity economies. This chapter will discuss certain agencies of change, namely: (1) the consulting firm, (2) the international corporation, and (3) multilateral AID and international agencies. An appendix to the chapter reviews some of the salient statistics of bilateral and multilateral AID, important for the analyses which will follow.

Transferring or Learning?

The "transfer of technology" is commonly used to connote the process with which we are here concerned. Yet, this term is misleading, for it is not "transferring" but "learning" that is at issue—the capacity of a society for learning and applying what it learns in order to produce more with the same human and natural resources. Different kinds of information are useful for and relevant to increased productivity. Different means and instruments and agencies exist for producing, acquiring, discovering, communicating, assimilating, and evaluating such information and for incorporating it into choice and decision, into practice and organization. Ultimately, technological advance depends on what those individuals who live and work in a particular society have learned, including what they have learned about how to act upon what they know. This is true for every society, rich or poor, far ahead or far behind.

No society is self-enclosed. Every society learns from or makes use of the learning of others. This learning takes place through trade, training, and travel.

The importation of products or equipment is a means of using the knowledge of others directly, and also of learning from others, for the imported product offers itself as a model for analysis, experimentation, and imitation.

Information infiltrates, and society learns through all means of observation and through all channels of interpersonal and interorganizational communication. It learns through the in-migration of people. It learns by having its own people go elsewhere to observe, be trained, and return.

Change Agents in the Private Sector

1. *The Consulting Firm.* The facts concerning the modern consulting firm have not been ferreted out for systematic analysis nor formulated as history. We can speak only as an occasional observer and participant.

Clearly, the free-wheeling professional, the "business doctor" with his ad hoc engineering, his time and motion studies, his tax dodges, operating at the interstices of the firm in the matrix of industry, has been with us for a long time. But, the consulting firm, as a complex enterprise that associates, institutionalizes, and integrates a range of skills—of publicist and pleaser, of salesman and engineer, of politician and researcher, of bureaucrat and promoter, of academic and organizer—and that merges, affiliates, conglomerates, depersonalizes into quite massive, supranatural organizations, is a post-World War II phenomenon. It evolved in response to a particular need, and its significance should be understood in relation to a particular lack that developed in Western societies after World War II.

Prior to World War II and the Great Depression of the 1930s, public policy reflected the laissez faire outlook of ideological liberalism. Government was organized not to *do* anything; literally, it was the rationale of government organization to limit and constrain rather than to facilitate and augment collective decision and political action. The collective power was diffused and enchained in order to preserve property power and market exchange from the threat of political meddling. Constrained as it was, the political authority evolved as a cumbersome and unfocusable system of collective choice coupled with a bureaucracy that was virtually without scientific or technological competence whose rigid organization made no provision for creative initiative nor for flexible response to changing social phenomena. The ideological constraints of laissez faire were shattered by the Great Depression. During World War II the political authority, perforce, became the decision center and control locus in organizing activities of the highest scientific and technological complexity. This was done in the United States and Great Britain by co-opting the services of scientists from the universities and of technologists, organizers, bankers, engineers, and managers from private business into a ramshackle but temporarily serviceable power structure superimposed upon the old governmental system.

After the war, that superstructure was dismantled, but the tasks of government did not diminish into their former insignificance. Quite to the contrary, the magnitude and complexity of the government's tasks increased. To the political authority fell the responsibility for full employment, price stability, economic growth (hence, as the sine qua non of economic growth for accelerating technological advance), the development of weapons systems, education, science, health and "welfare," space exploration, urban planning, the transformation of institutionalized racism, and promoting the economic development of the LDCs. These new responsibilities of the political authority required the highest degree of managerial and technological competence, a mastery over science, and a capacity for flexible organization and response. But, unfortunately, congressional or parliamentary processes were not equipped to initiate, implement, control, or to evaluate complex activity and organization, and the established bureaucracy had not acquired the competence needed for these great new tasks.

The consulting firm was a response to this lack. Sometimes the satellite of a government agency (like the RAND Corporation was of the Air Force) or a university outcropping (like Harvard Advisory Service or the Stanford Research Institute), it afforded a haven for refugees from impoverished backwaters of Academe, civil servants driven out by McCarthyite persecutions in the United States or by the end of colonial administrations in Europe. It combined the flexibility of a private business and technical expertise with politico-bureaucratic savvy and a social orientation; thus, it was equipped to do for public agencies what those agencies were not yet organized or able to undertake themselves. In technical assistance particularly, consultant firms have acted on behalf both of donor and of recipients in performing a range of economic and technical services. Flexible and available, they recruit and recombine skills of all sorts in the planning and development of transportation systems and traffic controls, hydroelectric systems and power grids, water and irrigation controls, land reclamation, harbor and port construction, industrial feasibility studies, public administration, police organization, and so forth.

The consulting firm is, nevertheless, of limited value as an agency of development. Its position is subordinate. Its powers are marginal. Its operators are transitory, ad hoc, without continuing involvement or final responsibility. Hence, its work is liable to be quick and shallow. It does not remain to give force to its advice, though ideas need their champion and a program needs its advocates to explain it, to adapt it in the face of objections, to correct errors, to persuade and persist and arm-twist and persist and persist. Without that, novel recommendations or those that call for significant change are neglected and forgotten. It is the consultant who learns. What he has learned stays under his hat; he takes it with him when he goes, leaving behind massive mimeographed reports that join other dust collectors on the library shelves. He contributes tools and artifacts (bridges, harbors, roads, dams). He builds for the developing society, but is less likely to build into the developing society the essential competencies required for self-growth and development.

2. *The International Corporation.* That national economies are increasingly linked not only by trade, but also by the organizational overlap of international companies creates an important new potentiality for cross-societal communications, for technological interchange and mutual learning.

Whenever its operations are plugged into the LDCs, the international corporation is uniquely equipped to be the teacher of advanced technology and the promoter of technological advances. Not only are they natural conduits for conveying technological information to the LDCs, they are also able to convey information to the rich and advanced, concerning the particularities and needs, the opportunities and the means of developing opportunities among the LDCs. But why should the international company be motivated to act thus as a two-way conduit of competence and information? Possibly because the prosperity of the international corporation is to some degree at stake, since political security, higher productivity, and more affluent markets in the areas where it operates would always be to its advantage.

The modern corporation is no profit-seeking individual but is a species of collective activity that, in common with other forms of collective activity, finds coherence in an ideological commitment; no less ideological when the commitment is to maximize corporation profits than when it is to render a public service. The corporate group also operates in a frame of exogenous pressures and constraints, which, acting upon the corporate entity from the outside as well as the ideological commitment driving it from within, can change. During postwar decades both have changed greatly. A clue to the shift in the ideological understructure is the extraordinary development of the private *foundation* as the eleemosynary arm of the corporation; such foundations have contributed more effectively in AID for LDCs than has the State Department and its agencies. Thus, in 1970 Grants in Aid by private voluntary agencies in the United States amounted to $578 millions, which was about the same as total grants *and loans* made in AID by the government of West Germany, and was considerably more than the total grants *and loans* for AID made by the governments of Australia, Austria, Belgium, Canada, Denmark, Italy, Japan, the Netherlands, Norway, Portugal, Sweden, Switzerland, or the United Kingdom; greater indeed than the total grants *and loans* in AID from Switzerland, Sweden, Portugal, Norway, Denmark, Austria, Belgium, and Italy combined.

Grants in AID by private voluntary agencies from countries that are members of the OECD's Development Assistance Commission are shown in Table 1.

Not only has ideology changed as a driving force from within, the corporation's framework of external constraints and pressures has changed as well. After World War II colonial territories and the old fiefs and passive satellites of Europe and the United States became politically sovereign, aggressive, suspicious and sometimes xenophopic masters in their own house. The international corporation had to adjust to the new condition.

Table 1

AID GRANTS BY PRIVATE VOLUNTARY AGENCIES, 1970

	Amount ($ Millions)	% of GNP
Australia	$ 15.7	0.05
Austria	3.6	0.02
Belgium	14.8	0.06
Canada	47.8	0.06
Denmark	3.0	0.02
France	2.4	*
Germany	77.8	0.04
Italy	(5.0)	(0.01)
Japan	2.9	*
Netherlands	(5.2)	(0.02)
Norway	(3.9)	(0.03)
Portugal	0.8	0.01
Sweden	25.2	0.08
Switzerland	11.8	0.06
United Kingdom	42.3	0.03
United States	578.0	0.06

SOURCE: Table is from OECD data released June 28, 1971, and covers only the members of the OECD's Development Assistance Commission (DAC). Figures for Italy, Netherlands and Norway are provisional. In general, it is considered that the grants are understated.

And, with time, the LDC also is acquiring a more sophisticated understanding of the potential value of the international corporation as a conduit for information and an instrument for learning, reflected in the injunction of the Pearson Report, that rather than requiring that a share in foreign subsidiaries be turned over to local ownership (which probably would mean no more than a giveaway of shares to the rich minority), "it seems better to press for things which give a greater assurance of gain, such as technical and managerial training of local personnel, assistance to local supplying industries, the establishment of a plant large enough to serve export markets, and limited tariff protection and tax concessions."[2]

This question remains: what can the international corporation be expected to do on its own initiative in realizing upon its potential as an instrument of economic development? It is not unusual to hear comparisons made between the assets, income, working population of the modern corporation, and the assets, income, and working population of some small sovereign nation. Both are complex groups, both are "governed." Both are productive. Perhaps the corporation is the wealthier and the more productive of the two. But essentially this analogy does not hold. It is like comparing a rabbit with an elephant's trunk. The trunk of the elephant can easily lift and crush the rabbit, but only if it is joined to the rest of the elephant. As a severed limb, it can do nothing. But the rabbit, viable as a closed system, can venture in, venture out, sniff, nibble, and scamper away. So the international corporation, large and powerful though it may be, operates only as a part of a larger system that is not of its doing or design—of law and security, of energy and transportation, of education and health care, and the open availability of a great range of information, services, a work force, and resources and supplies. Hence, whatever its own orientation and readiness, the international corporation cannot realize its potential contribution to economic development by itself, but only within and as part of a larger plan. The consciousness of the potential and the ideas of how to exploit the potential of the international corporation as an instrument of LDC learning and as a cross-conduit of information has hardly yet dawned in the ideological space of political or of corporate motivation.

Multilateral AID and the International Agency

AID is specifically the instrument of donors to change the technological capabilities of the LDCs. It can be organized bilaterally, donor to recipient, or multilaterally through the intermediary of an international agency. The record of bilateral and multilateral AID by the United States, Japan, and the West European countries associated with the Development Assistance Commission of the O.E.C.D., i.e., the DAC countries, is shown in the appendix to this chapter. Bilateral AID will be analyzed in some depth in the case studies of French, British, Dutch, and German AID to follow.

There are certain advantages in organizing science resources multilaterally, with an international agency serving as the planning and activating mechanism, given the following conditions: (1) information forthcoming from the research of many nations requires multinational arrangements for its effective transmission; (2) the problems susceptible to scientific attack are regional in character—e.g., locust control—and the region includes the territories of many nations; (3) the low-cost achievement of objectives will be served by the cross-national recruitment of scientific talent and pooling of research facilities.

Also, there are very good reasons why multilateral arrangements, in general, and international agencies, in particular, should be avoided. The perennial difficulty in organizing anything through an international agency is in the ambiguity of the mechanism of decision, a mechanism that is more ambiguous to the degree that there are significant differences in the policies, needs, values, and political and economic structures of those who participate, and to the extent that the agency itself lacks a specific mandate and a clear-cut assignment. International arrangements normally interpose another thick and obtuse bureaucratic layer into the apparatus of action, removing operational capabilities a further distance from decision and control, attenuating the links between choice and action, between action and responsibility, adding uncertainty, clogging communication.

AID can be organized through international agencies: (1) that have a substantial representation of both donors and recipients, (2) that are predominantly of technically advanced donors, or (3) that are predominantly of technically backward recipients.

The United Nations includes recipients and donors. Of the UN agencies organizing science resources for development, the most important is the Food and Agriculture Organization (FAO), which has recruited experts for missions related to food and agriculture, and has engaged in, or itself organized, science-based programs and projects.

ECAFE (Economic Commission for Asia and the Far East) has been the most successful regional organization in coordinating science-related activities for economic development.

The Commission for the Organization of African Unity, with its Commission for Scientific and Technical Research, was originally an organization of imperialist powers. As such, it set up effective (though small, low-powered) agencies for technological learning and research coordination throughout the continent, such as the Bureau Interafrican des Sols (BIS).[3] The founding nations were eliminated. Their place was taken by newly independent African governments, under whose control the agencies of scientific coordination seem virtually to have perished.

The activities of the World Health Organization (WHO) in the field of human health parallels those of the FAO in agriculture. The primary concern of UNESCO is in education, but it has also been concerned with science planning and policy.

The European Economic Community (EEC) and the Organization for Economic Cooperation and Development (OECD) bring together rich and technically advanced donor countries. Both are involved in AID programs.

The EEC, formerly known as the "Six," has as its primary task, to develop a common market in Western Europe. Nevertheless, through a "European Fund," EEC members contribute to development projects in former colonial territories. The OECD includes all the Western European countries, as well as Yugoslavia, Turkey the United States, and Japan. In an earlier incarnation, this agency helped to administer the Marshall Plan. More recently it has interested itself, among other things, in economic development. It has a "development center" where reputable scholars are handsomely paid to carry out allegedly relevant studies.

As a pressure group for the poor and a moral arbiter for the nations, the international agency can attempt to influence and propose criteria for the guidance of the national policy among donors. It can, and increasingly does, serve as a coordinator of donor activities and LDC planning in the field. It can also undertake to organize AID programs. Most often it has acted as a middleman, recruiting experts for work in LDCs, at the requests of recipients and with the support of donors.

In the instance of the UN family, the LDC asks for some special assistance, and the FAO, ILO, WHO, or UNESCO hires specialists (such as geologists, pedologists, statisticians, urban planners) and sends them to work in the recipient country, paying them in whole or in part from donor contributions. Residing in the LDC for a limited period of time, the expert conducts his survey and submits his report, or performs some operational task. This approach has its limitations:

(1) The expert is likely to be isolated in a new, strange environment. His assignment may well be terminated before he has discovered, or learned to manipulate those levers of power that are requisite to concrete achievement.

(2) Where the problems are familiar, and prescriptions are routine, then the visiting expert will suffice. But where new problems are confronted calling for complex analysis and creative solutions, then it is not the expert on his own, but the expert organization—the information-producing machine with its battery of specialists, its tools and logistical supports, its institutionalized ties with world science, and its

capacity for continuous research and long-range planning—that is required.

(3) There is nothing to assure a follow-up on what has been started by the visiting expert. If nothing comes of his efforts, if his recommendations are abandoned or ignored, no one will be called to account or questioned, nor is there any systematic means for him to evaluate the impact of his mission, hence to learn from it success or failure.

(4) Through confrontations with the problems of development, information will be accumulated and, through experience, skills will be developed. But when that confrontation is by the visiting expert who comes, reports, and goes, correspondingly, it is less likely that such knowledge and skills will become integrated into the system of choice and action. Nor is it likely that the visiting expert will have the time or the power to propel his proposals to the point of practical realization.

(5) A body of experts has been available for such assignments only as a consequence of decolonization and the disemployment of those who formerly had worked for colonial governments. Those former colonial civil servants are aging. Their ranks are depleted. The training that produced them has not been replicated for the following reasons:

  (a) The colonial system, with its limited but clear-cut objectives, permitted the long-range anticipation of professional needs, the development of a training system to supply those needs, and the offering of terms and conditions of tenure and promotion sufficient to induce the required flow of trained or trainable candidates. All this no longer exists.

  (b) AID, through visiting experts and ad hoc projects, allows no clear and definite anticipation of professional needs. Hence, government agencies have been unwilling to offer a career service for scientists and science-trained technologists specializing in the tasks of development.

  (c) Educational institutions in donor countries are often handicapped in such training, inasmuch as the links are lacking that would permit their students and their faculties to have direct contacts with and research experience in development-related environments.

(d) Educational planning and the provision of a career service for experts to work in AID programs is made vastly more difficult by the fact that predictions of the need for scientists for development programs can be made only in global terms, and possibly only on such terms can a plan be formulated and implemented for satisfying that need. But, if the problem is global, the responsibilities for relevant planning and action remain national—and uncoordinated. How many geologists should be trained in, say, Holland for overseas service? They will not be used in Holland nor in any Netherlands' possession. Their career opportunities will depend on the needs (and the national capacity to fulfill those needs) in Latin America, Africa, India, the Near East, and Asia. What will these needs be? And to what extent will multilateral or bilateral technical assistance support their fulfillment? The employment opportunity of the Dutch geologists would depend on the number of geologists trained and made available by other donor countries, as well as those trained in the developing economies themselves. Moreover, the possibility of ever achieving a dynamic equilibrium between the supply and the world demands for development-trained scientists and the possibility of providing scientists so trained with stable employment opportunities are greatly reduced by the barriers (both in language and in governmental practice) which block the cross-flow of personnel between AID projects under various national sponsorships. The Dutch geologists, for example (or for that matter, the Canadian, Australian, or Indian geologist), will *not* be hired, no matter what his qualifications, to fill some expert task in an AID project supported by the United Kingdom, no more than the English geologist will be hired for overseas geological surveys under German technical assistance programs.

Bilateral and Multilateral AID: Facts and Trends[4]

TABLE 2 COMPARES MULTILATERAL AND BILATERAL AID FOR DAC donor countries, ranked in the order of the size of their GNP. The relative magnitude of multilateral AID tends to be less, inasmuch as there are political links between the donor and a particular group of recipients, and

Table 2

MULTILATERAL AND BILATERAL AID, 1970

Country	GNP* ($ Millions)	Population (in thousands)	Multilateral as a % of Total Public AID†
United States	976.5	204,800	13
Japan	196.2	103,500	19
Germany	185.3	61,559	22
France	145.9	50,775	11
United Kingdom	121.0	55,812	11
Italy	92.7	54,459	49
Canada	80.9	21,406	23
Australia	34.3	12,700	6
Sweden	31.4	8,046	34
Netherlands	31.3	13,032	21
Belgium	25.1	9,700	23
Switzerland	20.4	6,300	40
Denmark	15.7	4,920	37
Austria	1,415.7	7,400	34
Norway	11.2	3,880	60
Portugal	6.3	9,670	8

*At market price and 1970 exchange rates. Most figures are provisional.
†Bilateral AID would be reciprocal.

22

Net Disbursements

Table 3
PUBLIC AID (NET) 1960–1970

Million U.S. Dollars

	1960	1961	1962	1963	1964	1965	1966	1967	1968	1969	1970
Australia	59	71	74	96	100	119	126	157	160	175	202
Austria	*	3	7	2	12	31	31	26	23	15	19
Belgium	101	92	70	80	71	102	76	89	88	116	120
Canada	75	61	35	64	78	96	187	198	175	245	346
Denmark	5	8	7	9	10	13	21	26	29	54	59
France	847	943	976	852	828	752	745	826	874	955	950
Germany	237	330	398	393	460	430	440	528	553	595	599
Italy	77	60	80	70	48	60	78	155	146	130	152
Japan	105	108	85	138	116	244	283	384	356	436	458
Netherlands	35	56	65	38	49	70	94	113	123	143	196
Norway	5	7	7	10	10	11	14	14	27	30	37
Portugal	37	44	41	51	62	21	24	47	35	58	31
Sweden	7	8	18	23	33	38	57	60	71	121	117
Switzerland	4	8	5	6	9	12	13	13	24	30	30
United Kingdom	407	457	421	414	493	472	486	485	413	431	447
United States	2,702	2,943	3,181	3,567	3,576	3,400	3,394	3,497	3,228	3,092	3,050
Total DAC Member countries combined	4,703	5,198	5,471	5,812	5,955	5,872	6,070	6,618	6,325	6,625	6,813

inasmuch as the donor economy is large. The reason for this will appear in the chapter on the Netherlands.

Table 3 shows the changes in Public AID (Official Development Assistance) by DAC members between 1960 and 1970. In reading the data, account should be taken of the considerable and continuous inflation—hence, depreciation in the real value of the dollar—during this period. Thus, in real terms, American and British AID have sharply declined from 1963 onward. AID has sagged in France, but has increased in Germany and the Netherlands.

Figures 1 and 2 and Table 4 show the relative position in 1970 of donors in relation to total net flows and Public AID to recipients. The outstanding contribution of France and of the Netherlands relative to the others is to be noted.

Table 4
NET FLOWS AS A PERCENTAGE OF GNP

	Public AID Development Assistance		Other Official Flows		Private Flows		Total		
	1969	1970	1969	1970	1969(a)	1970(a)	1969(a)	1970(a)	1970(b)
Australia	0.56	0.59	—	0.02	0.1	0.46	0.74	1.07	1.12
Austria	0.12	0.13	0.01	*	0.50	0.51	0.65	0.65	0.67
Belgium	0.51	0.48	0.01	*	0.61	0.69	1.12	1.17	1.23
Canada	0.34	0.43	0.07	0.07	0.09	0.22	0.50	0.72	0.77
Denmark	0.39	0.38	*	-0.02	0.69	0.24	1.08	0.60	0.62
France	0.68	0.65	*	0.02	0.54	0.57	1.22	1.24	1.24
Germany	0.39	0.32	-0.03	0.07	0.98	0.37	1.33	0.76	0.80
Italy	0.16	0.16	0.01	0.02	0.86	0.60	1.03	0.78	0.79
Japan	0.26	0.23	0.22	0.35	0.27	0.34	0.76	0.93	0.93
Netherlands	0.51	0.63	0.02	0.02	0.78	0.77	1.31	1.41	1.42
Norway	0.30	0.33	0.08	—	0.38	0.23	0.77	0.56	0.59
Portugal	1.04	0.48	0.37	0.42	0.32	0.11	1.75	1.01	1.02
Sweden	0.43	0.37	—	—	0.30	0.28	0.73	0.65	0.73
Switzerland	0.16	0.15	-0.03	-0.02	0.51	0.48	0.64	0.61	0.66
United Kingdom	0.39	0.37	*	*	0.65	0.65	1.04	1.02	1.06
United States	0.33	0.31	0.02	0.02	0.16	0.22	0.51	0.55	0.61
DAC Member Countries Combined	0.36	0.34	0.03	0.06	0.35	0.34	0.75	0.74	0.76

(a) Excluding grants by voluntary agencies
(b) Including grants by voluntary agencies

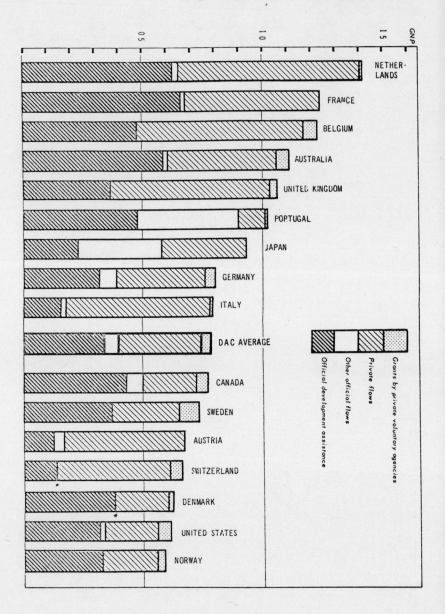

Fig. 1–1 Total net flows related to GNP (1970).

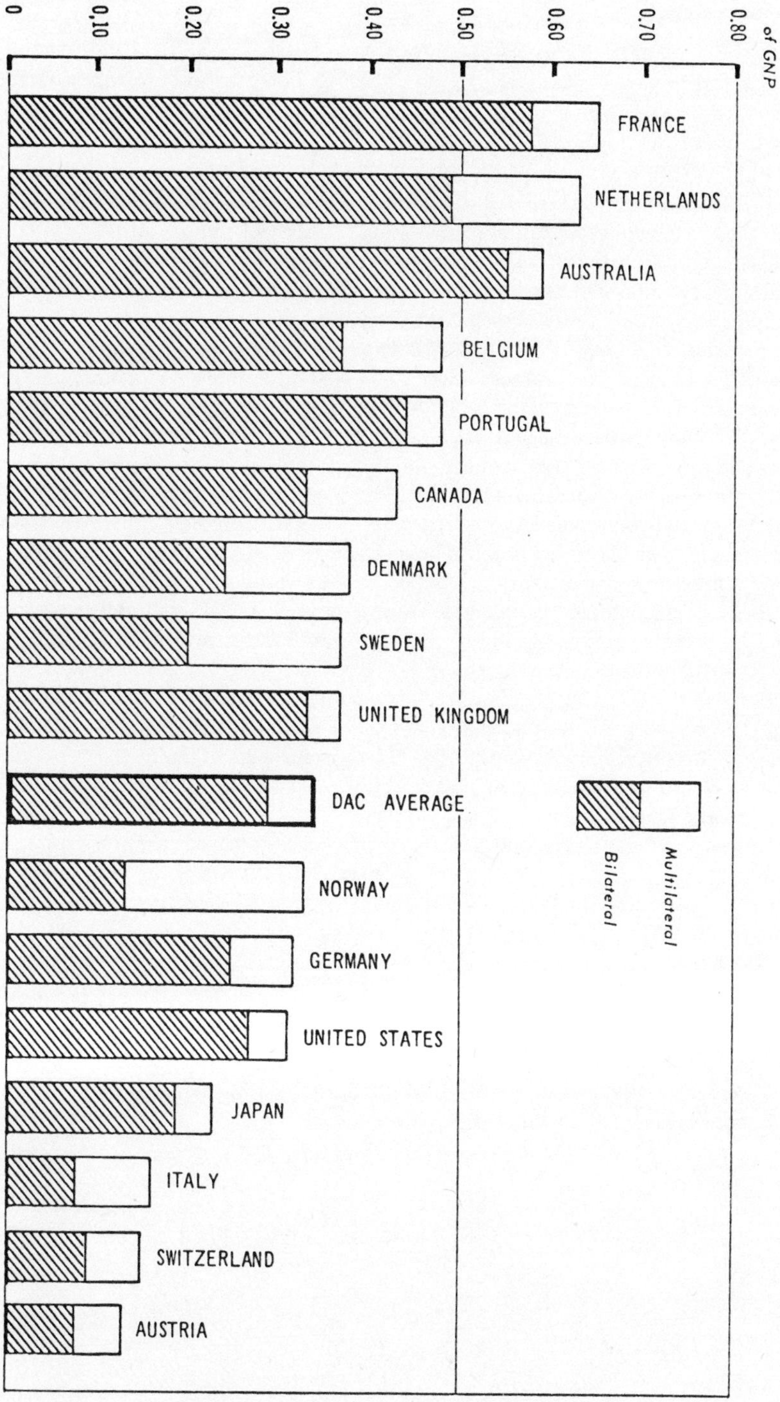

Fig 2. Public AID Related to GNP (1970).

NOTES AND REFERENCES

1. Lester B. Pearson, *Partners in Development.* (New York: Report on the Commission on International Development, United Nations, 1969).

2. Pearson, *Partners in Development,* p. 112.

3. In 1964 the BIS had been operating for sixteen years, with its center in Paris. It was a very low-budget institution, consisting of only two soil scientists, two documentalists, one translator, three typists, and the operator of a mimeograph machine. With this minimal staff, it maintained a specialist library of 1,276 reference works. It undertook continuously to match the current outputs of scientific literature with the research interests of all scientists working in Africa on soil-related problems. It regularly reviewed 242 journals dealing with soil science and agronomy, and recorded article references on index cards. Following an extremely detailed classification plan, these cards were distributed according to the specific research and technical problems to which they related, and, two or three times each month, batches of cards were forwarded to individual researchers and research organizations in Africa (or elsewhere on request) according to research problems currently being undertaken there, or to the problem area on which researchers have been asked to be kept informed. Copies of articles and documents were reproduced and forwarded to the field researcher at their request. These services, at a very small additional cost, could have been extended from Africa to field researchers on problems of soil and agronomy throughout the whole developing world.

Further, the BIS published an important quarterly review *African Soils,* which provided an outlet for the publication of regional research results and for scientific papers delivered at African conferences. It also published a *Monthly Analytic Bulletin,* which yearly analyzed from 230 to 300 articles considered to be of particular significance for soil science and soil conservation.

At least once during the year, one of the BIS soil scientists visited every research station in Africa concerned with soil problems. Thus, the BIS acted as an informal agency coordinator for liaison and research between far-flung, multinational research activities.

The BIS itself undertook library research on request, and acted as a liaison between field scientists in Africa and specialized research agencies in Europe and the United States.

By 1971 that small science agency and presumably the others like it had vanished.

4. Unless otherwise specified, data is from OECD, "Development Assistance in 1970 and Recent Trends," 28 June, 1971.

Chapter 3

Forms of Science-Based AID

HOW CAN SCIENCE-TRAINED MANPOWER IN THE UNIVERSITIES, IN RE-
search, and in development organizations or elsewhere—or the competencies they have acquired or the information they produce—be used to support economic development?

Replicating Technology

In the simplest case, the transformation of technology requires that
equipment and plant design which have operated elsewhere be duplicated
and transplanted in a low-productivity society, as when diesel locomotives
are brought in to replace steam locomotives. Such a problem is merely one
of procurement, and the role of science is peripheral.

Not all technologies can be so easily transplanted. A technology that is
eminently workable and highly productive in one society may be ill-adapted
to another. What are considered as "advanced technologies" have generally
evolved in a social and physical environment which differs significantly
from that of the LDCs. Correspondingly, such technologies are likely to be
ill-adapted; indeed, their use in the circumstances of the low-productivity
economy may be entirely uneconomic and technically retrogressive. Characteristically, advanced technologies have developed in temperate climates,
whereas LDCs are, for the most part, located in tropical and subtropical
zones. For that reason, different vegetation, different fish and fowl, different
animals flourish in each. The structure of the soils and the practices appropriate to soil conservation will differ. There are different crops, differently
cultivated, with different problems in preservation and processing. Different
diseases attack men, animals, and plants. When the diseases are the same,

29

their vectors are likely to differ. Correspondingly, it is not possible to transfer the technologies or the techniques and information associated with the sciences of agriculture, horticulture, animal husbandry, medicine, and public health developed in the temperate conditions of high-productivity societies directly to low-productivity societies in tropical and subtropical zones. As in the physical, so also differences in the social and economic context of technical operations, hence differences in the relative costs of inputs, may bar the direct transference of advanced technologies.

Science-Trained Manpower for Business Choice and Political Control

Given "advanced" technologies, and elements of technologies that may or may not be optimal or operable in the LDCs, the judgment must be made as to what is usable and what is useless, as to what is better, as to which is best. Such judgment is likely to require a high degree of skill and knowledge. Where the technology is science-based—e.g., a data-processing, metallurgical, or pharmaceutical operation—the judgment of feasibility would require science-trained capabilities; and science-trained manpower would be needed to install and organize the operation and to keep it competitively abreast through research. Hence, even where it is possible to advance technology simply by imitating what is proven and practiced elsewhere, science resources may be required as participants in industrial (or business) choice and control.

The Science Resource in the Adaptation and Redesign of Technology

Thus, even when transformation can be achieved through imitation, it nevertheless will, to some degree, rely on the science resource. Often imitation is not enough. Sometimes that which has been successfully established and practiced elsewhere must be modified to suit the special circumstances of the developing society. Correspondingly, as contexts differ, the necessary adaptation will be more complex and difficult.

The greater the differences in the context of operations, the more difficult will be the process of adaptation. This, incidentally, may account for the paradoxical emphasis on industrial development by predominantly agricultural, low-productivity societies; since the ineradicable differences in the natural parameters of agriculture (and also the very deep differences in the social circumstances of agriculture) in the LDCs, as compared to high-productivity societies, reduce the possibility of simple imitation and make agricultural technologies more difficult to adapt than industrial technologies.

What sort of difference in the contexts of operations make the adaptation and redesign of technology necessary? Labor skills may be lacking. Transport, communication, power facilities, and other components of the infrastructure are likely to be of another order. The size of the available market, the pattern of relative factor costs, and output priorities are likely to be different. So also is the institutional framework of individual or collective choice. What about the capacity to modify and adapt technology to a very different operating context?

Sometimes no more is called for than on-the-spot adjustments of imported equipment or modifications of a conventional plant. But a piece of equipment and a process design are expressions (partial expressions, partial embodiments) of a body of knowledge. Inasmuch as technologies are science-based, their readaptation, indeed their re-expression under altered parameters of choice, will, correspondingly, require a knowledge of the relevant science.

Development-Oriented Science and Scientists

The knowledge of science and technology accumulated in advanced economies will be largely irrelevant to the different problems encountered in developing societies. Different information is needed. And if the information is not there? Then it must be produced. And it is the capacity to produce such information—which is the very function of scientific research and the capacities that go into such research—that is required. What is, then, at issue is not technologies practiced and proven elsewhere, or even the corpus of information underlying those technologies, but, rather, the research and problem-solving capability itself, i.e., a development-oriented science.

Thus, the science resource is required to evaluate and replicate technology, to adapt and redesign technology, and for the development-oriented research to solve the special problems and to provide information for the evolution of novel high-productivity techniques suited to the circumstances of developing societies. We can illustrate all this with a single example. Suppose the need is to control insect pests on a tropical plantation. Spraying equipment must be imported. The proper choice of the equipment requires technological capabilities. Such choice is at the level of evaluation and replication. It may be necessary to adapt equipment to the climate, the terrain, the shape of the plant to be sprayed, the locus of its infestation, and the skills of indigenous labor. Such adaptation requires another level of technological, even a science-trained, capability. It might be necessary to compound an appropriate insecticide. If existing knowledge offers no effective means for controlling the pest, new information on the habits of the insect or the pesticidal efficacy of available materials will be needed. Such information will have to be produced through development-oriented research.

Scientific skill and knowledge are not simply auxiliary to the functions of planning and innovation. It should not be supposed, as it too often has been, that it suffices to have scientists "on call," available to the businessman, politician, or bureaucrat. On the contrary, the mastery of science itself affords avenues of development planning, for organizational innovation and for policy initiative that are closed to those who have not that mastery. The skills and knowledge of the development-oriented scientist can be used in policy making and in the planning and programming of economic development.

Creating a Hospitable Context for Advanced Technology

It is not only a matter of adapting technology or of creating a technology adapted to a particular context of operations; alternatively, the context may be adapted to the available technology. New and complex techniques using concrete as a material in residential construction have been developed for certain low-productivity economies because a lack of skilled craftsmen precluded the technology of construction conventionally practiced elsewhere; alternatively, carpenters and bricklayers could have been trained. New and complex techniques of aerial survey for mapping and inventorying of natural and social resources have been developed more or less specifically for the low-productivity economies; alternatively, roads and internal transportation facilities could have been built which would have made it possible to use conventional survey methods.

More fundamentally, the whole environment of the LDC may be transformed to facilitate the processes of learning, adaptation, and technological transformation. This surely is the purpose in upgrading the infrastructure, in the building of roads, harbors, airfields, and in making available transport facilities, in providing protection against contagion, violence, theft, fraud, flood, hurricane, in offering the services of banks, equity markets, and so forth. Especially relevant to the capacity to adapt technology, or to adapt to advanced technologies, are those science-based components of the infrastructure which raise the capacity to control, adapt, and innovate in any sphere of technology. This science infrastructure requires science-trained manpower.

The Science Infrastructure

The word *infrastructure* has been used to describe the context of facilities, institutions, and organized conditions within which enterprises can develop, and government agencies can plan and direct, forms of economic activity. A physical infrastructure, is embodied in things such as roads, harbors, dredged rivers, and canals. There is a social infrastructure, including the law courts and the police to protect property and to enforce contract. And there is also a science infrastructure to provide services and facilities that help decision makers test, measure, and quantify the variables related to their choice; that provides them with quick accurate information concerning change in the natural parameters of choice, that enable them to determine exactly and to specify precisely the performance characteristics and requirements of the materials, products or structures that they intend to produce or plan to use, and to forecast the future parameters of political policy or business decision or the likely consequences of alternative choice and action. The science infrastructure is an especially important condition for the successful establishment of the science-based technologies.

Consider, for example, a single component of the science infrastructures, namely, the capacity to measure and to test. This capacity, (which is embodied in commercial laboratories and public services, in simple tools and in highly complex equipment, in mechanical skills and in research capacities, and in an accumulated complex of experimental data) is essential to quality control, to standardization, to technical specification and communication, to procurement (as a means of determining the comparative characteristics or suitabilities of the available alternatives) to innovation (in prototype testing), to sanitary control, to building codes, to food inspection, to a diagnosis of the breakdowns of industrial processes. Here is a structural form: what stresses can it bear? Here is a metal alloy: how rigid is it? how brittle? how much pressure will it stand? how much heat? Here is a new plastic: what are its characteristics, and hence where, with its particular balance of values and defects, might it best be used? Here is a paint: how well will it protect against corrosion? Each of these questions can be answered only through a capacity to test and to measure. In no society in the world is the capacity to test and measure wholly adequate. Nevertheless, this capacity, which should be shaped to a particular complex of activities and to a particular level of development, is absolutely prerequisite to modern industrial organization. It is but one element of the science infrastructure. Others would include the capacity to provide scientific and technological information and data through libraries and documentation centers, or the capacity to analyze soils, minerals, and ores, or to process data, or to provide statistics and statistical services.

In advanced countries the science infrastructure has evolved so gradually, that one is hardly aware of it. In building the science infrastructure requisite for economic development from scratch, it becomes (for the first time) necessary to recognize the component elements of that infrastructure and to understand their functional values and their costs. What service is to be provided? What infrafunctions must be built in? If it is capacity to measure and test, then what is to be tested and measured? And for whom? Is it the quality of textiles, the impurities in rum, the pollution of rivers, the characteristic of plastics, the purity of foods, the strength of metals, the quality of fibers? Is it to be for an agricultural extension service, an inland waterways bureau, a sanitation inspectorate, a private industry? If there are many areas where this measurement and testing capacity would be useful, what is the order of their priority? Only based on such information and judgments can the appropriate science infrastructure, as an institutional back-up to choice, be created.

By institutional arrangements, by a chain of responsibilities and powers, the infrafunction must be linked into politics and practice, into development programs and plans, and into private choice. In the LDC it may be necessary, not only to create the service but also to construct its linkages into practical choice and action. And the infrafunction must also have the (research) capacity to adapt itself and to evolve. This requires viable links to relevant fields of world science and technology. Thus, institutions for test and measurement must have a two-way tie: one to assure that information finds its way into economic choice and the other to enable the institution for test and measurement itself to learn about and to benefit from relevant scientific and technical advances throughout the world.

Knowledge into Action

The scientist may produce information relevant to political and economic choice, but such information need not come to the attention of those responsible for policy and practice; or if it does, they need not understand it; or if they do understand it, they need not be motivated to use it; or if they are motivated to use it, they need not have the skills or resources to take the action which is indicated. To create a viable institutional linkage between the scientist and the decision maker may be the most difficult problem of all.

Agricultural extension services, for example, intended to transmit the results of agricultural research into better farm practices, have been established in developing countries. The organization of the extension services may be the same as in developed countries—but they need not work at all as they do in developed countries. Extension services in the United States or in Western Europe are designed to serve the farmers, and farmers have been an effective and demanding clientele, with lots of political muscle, on whose support and patronage the continued existence of the agricultural research establishment and of the extension service depends. The extension service makes available to the farmers a battery of specialists. A farmer asks the questions, and the specialists supply the answers. The farmer evaluates those answers in terms of his needs and his situation and puts the bits together into a coherent plan of investment and transformation. What happens when the same battery of specialists confronts the Asian or African peasant, farming by ancestral rote? Can that peasant exert on the extension service those pressures that shape the system to his needs? Before the specialists, he is passive, indifferent, or inactive, unable to formulate problems, ask question, demand answers, or to evaluate the information he is given. He is offered information, but he cannot integrate it into a coherent plan of action and change. The extension service is designed to support the active, aggressive politically potent Western farmer in his effort to transform his farming practices. But the peasant is uninitiated into the venturesome and risky task of transformation. And the extension specialist? Without a dialogue with the farmer, without a critical and demanding clientele who will keep his nose to the grindstone? He becomes an academic researcher looking for recognition from his academic peers. The elaborate servicing mechanism turns out to give no service.

Technology and Forms of Social Organization

Technology implies organization. Ordinarily, one thinks in this connection of physical forces and materials and machines as the things to be organized, but there is also always the organization of men and motivations and values. And every technology is embedded within social organization and often is an instrument of social organization. For this reason the feasibility of replicating a technology operational elsewhere will depend on the degree to which the forms of social organization, including the forms of political, educational, scientific, industrial organization, are similar or congruous.

For example, computer-based data processing constitutes one of the most spectacular components of the science-based technologies of Western societies. But computer-based data processing is more or less specifically a support in, and is important only with respect to, the planning and control functions of very large public or private organizations. Thus, computer-based data processing has hardly any significance in the small enterprise or agricultural sectors of the American economy, or wherever decision making is decentralized, and when operations are carried on in the discrete entrepreneurial entities of price-competitive markets.[1] Therefore, the feasibility of transferring computer-based data processing into developing economies presumes highly complex planning and control functions by large private or public organizations. If centralized control through large public or private organizations is not a characteristic of the economy of the developing society, then there will hardly be any place in it for the computer. Of course, the emergence of computer-based data processing—since it does increase the potential effectiveness and does lower the potential costs of centralized control or programming—makes large-scale operations and centralized planning more attractive as organizational alternatives. But whether a society will (or should) choose to bring more activities within the range of centralized political direction, or within the scope of large corporate enterprise, will depend not only on whether that society has the science resources to operate the computer-based technology effectively, but to a far greater extent, on the cultural proclivities and social values prevailing there.

The Structure of Cognitions

The social milieu that provides a welcoming environment for science-based innovation, and that determines its capacity to assimilate advanced technology, depends upon, and is determined by, far more than the existence of the facilities and the availabilities of the services of the science infrastructure. It depends upon, and is determined by, what those who constitute a society as a whole or in its parts *know,* what they *want to learn,* and the *capacities* they have acquired *for learning.* We shall call that knowledge; and its related orientation for and capacity to learn, "cognitions." Before a judgment can be made about what is lacking and before a learning program for a developing society can rationally be devised, there must be a conception first as to the cognitions that are needed for that society to respond to (and avail itself of) learning opportunities that inhere in accumulations of scientific skills, in science-based knowledge, and in advanced technologies throughout the world.

There are, of course, those cognitions that constitute the *capacity to communicate,* intragroup and intergroup, the capacity for dialogue, question and response, criticism and argument. Whatever its pietistic or authoritarian overtones, the establishment as a universal practice in postrevolutionary Chinese society, of daily discussion of the words and thoughts of Mao Tse-tung (or of anything else) must eventually have an immense effect upon the capacity of the common man to express himself, and for groups to communicate together systematically. That capacity to communicate is a prerequisite for all learning.

There is also a pyramid of cognitions directly related to the capacity of a society to respond to technological opportunity. At the base is the *cognition of mechanism,* an understanding of the machine, of its logic and its limits, of the demands it makes upon those whom it serves and who serve it, and an appreciation of its symmetries and power. Emerging out of the cognition of mechanism, carrying it into vocation and invention, is the *cognition of technique.*

A mass cognition of mechanism at the base and a middle mass of mechanical and technical skills at the center give to any society its capacity to respond to the signals of technical leadership and, also, spontaneously to adapt to the requisites of a particular technology, and to adapt technology to the particular circumstance and the cultural proclivities of social groups.

Emerging out of this middle mass of technical skills, embodying but transcending the cognition of technique, is the capacity to put the skills together, to organize men and machines, materials, and information into complex and viable processes. This is the *cognition of process,* that evaluates, allocates, integrates, that sets the pace and rhythm of work, that harnesses and maneuvers motivations,—evades conflicts, gives direction, sensitized to purpose and cost as the essential underside, the perpetual shadow reality of all effort and organization,—that perpetually engages itself in the search for, and in the design and evaluation of, output, input, and organizational alternatives.

At the apex of the pyramid, interacting with the cognition of process, is the mastery of science. Development-oriented scientists and science-trained engineers link the world cognitions of advanced technology to the locally focused cognition of process, and they bring to bear the research competence in providing new information relevant to choice and transformation.

Which is the key cognition? The cognition of process, surely. This is the pivot around which all the rest must turn. The cognition of process is not the same as the outlook and skill of the self-seeking, wheeling-dealing, price-manipulating entrepreneur in economic theory. In Latin America, for example, there are entrepreneurs aplenty, but the cognition of process is rare, and so is technical advance and economic progress. The cognition of process indeed merges with and emerges from out of the cognition of mechanism and the cognition of technique. Only the cognitions of science stand apart as something created by artifice and design.

The science cognition may be at the very apex, but the base needs to be built before the apex. With or without the scientific elite, an economy where there exists the cognitions of mechanism, of technique, and of process can propel itself powerfully forward on the path of economic growth. And without those cognitions, the scientist and his knowledge and skills will have no significant economic impact.

In Western Europe and the United States the cognition of mechanism, the skills of the mechanic and technician, and the cognition of process have generally been generated outside the scope of formal education. They have evolved as the product of a particular history. They have been passed on through commonplace experience and day-to-day observation in already industrialized societies. They are recreated and extended through ad hoc training and apprenticeships. The cognition of process particularly is absorbed through experience as a participant in established processes at various levels of managerial and technological responsibility. But, in low-productivity, preindustrial societies the mass cognition of mechanism and the cognition of process cannot be acquired spontaneously through day-to-day observations and experience. Nor does there exist that critical mass of technical skills able spontaneously to renew itself through communication, through ad hoc training to extend itself, through mutation and learning at the margin to transform itself. In the LDC, these required cognitions must be structured-in, systematically, by plan. The task becomes one for some kind of formal education.

How, for example, to inculcate preindustrial masses into a cognition of mechanism? How to cultivate a cognition of process when the experience of technological organization is not immediately within reach? It is doubtful that conventional schoolroom teaching can suffice. Radio and television can be powerful instruments, but only if they are integrated into an intensive and disciplined learning process. The educational experiences and the means of cognitive transformation developed in China, Japan, and Russia are highly suggestive. Even the old-time country fairs, which long ago brought new methods or products or equipment to the attention of rural America, offers a useful prototype—not only in rural areas for the sake of the peasant, but also as pilot demonstrations and expositions of technology for the city worker and entrepreneur. Surely the LDC should question the traditional European-American sequence in organizing its own educational system. In the traditional sequence, the student is first given a literary and cultural "base." He then proceeds to learn science in its more fundamental or "academic" aspects. Finally, he may be taught the "principles" supposedly underlying some selected sphere of technology. Only after he leaves the system of formal education does he focus on the skills and specifics of application. That system produces what it is intended to produce, an elite leadership, presuming a society that is capable of responding to the signals of that leadership. It relies on informal process of ad hoc and informal learning through apprenticeship, observation, and experience to produce the mass cognition of mechanisms, the middle cognition of technique, and the high cognition of process. In the developing society that presumption cannot be made. For that reason, perhaps the sequence of formal education needs to be reversed. The first objective then would be to introduce the whole youth of each new generation—indeed the whole society—into the elements of basic communication and the rationale of the machine. Out of the mass so trained, selected numbers would be taught mechanical skills and specialized techniques, and the mathematics and general knowledge required for these. Among those who had demonstrated the requisite abilities at the level of operating technique, the best would be selected out and through experience, observation, and guidance, helped to acquire the key cognition of process. And among these, those who could demonstrate particular talents for abstraction and disciplined articulation of ideas would be brought to the study of advanced mathematics and of technology-related science. This sequence is close to that installed in China following the cultural revolution, where the acquisition of the basic skills of literacy, logic, and communication are paralleled by universal inculcation of the cognition of mechanism and a general training into the cognition of techniques. All secondary school graduates there must, for an extended period of years, engage in, and presumably fulfill themselves by acquiring a mastery of complex organizational processes in industry, in agriculture, or in the army. Of these working cadres, some will be elected by their working comrades to study in the university, with all training and all scientific research in the university oriented toward practical action, social problem-solving, and technology.

Such an educational sequence will hardly produce distinguished academicians or Nobel Prize winners, but it may produce the structure of cognitions that will enable a social system to raise itself out of the ancient muck of hopeless poverty.

Science Planning, So-Called

In recent years, following the fashion, governments and international agencies have established teams for "educational planning" and then for "science planning" in the LDCs. This "planning" has, in fact, been primarily, or entirely, engaged in predicting the outputs of trained manpower, broken down by level and type of training allegedly needed in order to sustain a proposed pace of economic development. The underlying conception is simple enough. High-productivity economies (Germany, France, Great Britain, or the United States) have certain ratios of formally trained manpower (engineers and scientists in various specialties, lawyers, accountants, etc.) in proportion to their total population. Presumably that trained manpower reflects the real needs of a technically advanced, high-productivity society. Therefore, underdeveloped, low-productivity societies must have the same ratios of manpower trained in those branches of science and technology, if they are to become technically advanced. Their educational systems must be organized accordingly. Or if there is an economic plan, educational planning must be related to the plan. Suppose a steel industry is planned. Then education must seek to produce trained manpower as a ratio of the size of the planned steel industry equivalent to the trained manpower employed in the steel industries of the technically advanced countries of the world.

There are grave weaknesses in this kind of projection—weaknesses, that is, in taking the patterns of formally educated manpower in the advanced countries as models for educational policy in developing societies. As already suggested, in focusing on the formal education of elites, it overlooks the essential cognitions of mechanism, of technique, and of process which, in advanced economies, are acquired through apprenticeship, observation, and experience but which, in developing societies, must be inculcated through formal education. It fails to take into account the fact that the need for types of formally trained manpower is not only a function of technology but also is a function of the form of economic organization. When an economic function is organized through the decentralized individualized choices of many discrete entities in a price-competitive market, rather than through centralized planning or through large corporate aggregates, correspondingly fewer trained engineers as a proportion of the working population, and hardly any trained scientists will be employable in industry. Thus, as the United States was transformed by the so-called "organizational revolution" from a highly productive economy of small competitive enterprise into a highly productive economy dominated by enormous corporate organizations, there was a vast and rapid increase in the numbers of accountants, engineers, and scientists as a ratio of the working population. For example, between 1910 and 1960 in the United States, population about doubled, but B.A. and first professional degrees increased tenfold, and engineering degrees increased almost fortyfold.[2] It appears, moreover, that the employment of engineers and scientists is concentrated in massive corporate enterprise and in government. Thus, a 1959 survey found virtually no scientists or engineers employed in the more than two million smallest firms in the American economy. Of the 47,000 larger firms, the largest size category with 5,000 or more employees, employed fifty-two percent of all engineers and scientists in American industry. The employment of scientists and engineers among these largest firms, as a ratio of their total employment, was double or more that of any other size category in American industry.[3]

And finally, a substantial part of the scientific establishment in advanced countries is quite unrelated to technology or economic growth, ministering to man's esthetic or spiritual, rather than to his economic, needs.

We have spoken about the ways in which science-trained manpower and research organizations are needed or can be used in the processes of economic development, and about the conditions, cognitions, and social organization that are prerequisite for the effective use of science resources for the purpose of economic development. We have not yet spoken, however, about the ways in which donor nations use their science-trained manpower and research capabilities as a part of their AID programs, nor of the values from the point of view of the donor of science-based AID.

The Donor Use of Science Resources in Technical Assistance

Our concern is with the donor organization of science resources in AID programs. In fact, donors have organized their science resources for purposes of AID in, at least, the five following ways:

1. *The Inquiries Bureau and the Information Center.* Institutions have been created and maintained in the donor countries to make information avail-

able on request, e.g., information centers, identification services, libraries, experts available for consultations and advice, presupposing that there are those in developing countries who can ask the right questions and who can effectively translate responses into practical action and that the generally available store of science information will suffice for the needs of developing economies. These conditions need not be met. Frequently, development demands new sorts of information. And those who man the industries or control the governments of the LDC may not be able to ask the proper questions of the scientist or to translate the information the scientist offers into policy and action.

2. *The Visiting Expert.* Donors make experts in some field of science or in some science-based technology available to a recipient nation at its request. In contrast to the inquiries bureau, the visiting expert does not simply answer questions: he can formulate them. He can examine conditions on the spot and evaluate the relevance of available information in terms of these conditions. When a problem cannot be resolved within the scope of existing knowledge, then occasionally the visiting expert can dig up new facts, even adapting analytic methods and science-based techniques to the special conditions with which he is confronted. His recommendations, beyond the science-based information on which they are based, may include the means for their practical implementation, taking into account the play of personalities, the effective motivations to policy implementation, and the economic and social structure.

Nevertheless, as was suggested in discussing the work of the UN agencies, the use of the visiting expert has grave limitations and disadvantages. The effectiveness of science as an analytic and problem-solving technique depends not only on the personal competence of the individual scientist, but also on the institutional framework within which he operates and the organization of which he is a part, e.g., on the complementary competencies of his colleagues, on the availability of experimental facilities and research tools and libraries, on the organized flow of information concerning current or prospective research, on the conferences, seminars, and private interchanges by which the individual relates himself to a universe of thought and experience, and on those established contracts and ties through which the ideas and findings of the scientist infiltrate policy and produce innovation. As a visitor, the expert is pulled out of that framework. He acts alone, and no longer as one of the focal points of an institutional complex which, in turn, relates to and brings to bear the prowess of a world scientific community. If his visit is brief, he can hardly assimilate the pecularities of the social environment nor organize long-range research nor learn the ins-and-outs of administration, nor acquire that political know-how which always and everywhere is needed to transform information and ideas into policy and action. And if he stays for a long time, correspondingly, he isolates himself from the dynamism of his discipline. Ordinarily, he makes his report without subsequent opportunity to follow through on those hypotheses which are incorporated in his recommendations and to learn how right or wrong he was from subsequent success or failure. It is, or should be, the virtue of the science enterprise that its informational base accumulates and its potency increases as a consequence of individual studies, hypotheses, experiments, practical applications, and through all the failures, anomalies and partial successes which occur and the incidental techniques which evolve along the way. But this requires that there be the opportunity not simply of proposing hypotheses (implicit in all recommendations) but also for applying them and evaluating their results and for subsequently modifying and reformulating those hypotheses in a *continuous* confrontation with sets of related problems. And, to capitalize on these efforts, the results of this experience must systematically be "fed back" into the literature, into the consideration and interchanges of world sciences, into organizational learning, or into both. Science AID, through visiting experts, provides neither for continuity in research and application nor for a feedback of experience into the evaluation of developmental policy or development-oriented sciences.

3. *Support for Development-Oriented Research.* The donor can support research in its own institutes on problems of special concern to developing countries. Thus, deepening the information base relevant to development needs, at the same time training scientists and technologists expert in fields related to development.

Development-oriented science must evolve through an encounter with the problems and circumstances of low-productivity regions and, frequently that encounter requires research in the environment of those regions. And the competencies acquired through training and the information produced through research need to be tied into policy choice and practical action.

This double relationship, that donor science might evolve through experience in developing societies, and feed into action-programs or decision-processes there, would require that the science capabilities of the donor be systematically related to project AID or to development planning and programming in the recipient countries.

4. *Science-Based Projects.* Science resource may be engaged in discrete tasks in the recipient countries, e.g., to take an inventory of forests, to build a dam and develop a system of regional irrigation, to set up a materials-testing institute, to carry-out a geological survey.

The ad hoc project, as a form of science AID, has potential advantages. Donor countries can muster the competencies they have available and use them where they are most likely to be fruitful and, by so doing, strengthen their own scientific capabilities.

Science AID through ad hoc projects also has intrinsic disadvantages. It is likely to mean a number of abortive starts, without appropriate follow-through. It fails to provide for the continuity of experiment and for the feedback of experience required for the evolution of development-oriented disciplines. Nor can a system of ad hoc projects easily be reconciled with: (1) a rational plan for training development specialists and providing them with stable careers in the donor country or (2) with integrated development planning in the recipient country.

5. *Extending the Donor Research Capabilities into Developmental Planning and Programming.* Finally, the donor may gear the organization of its R and D agencies into the problem-solving apparatus and into development planning and programming in recipient countries. Only in the instance of French science in relation to economic development in French-speaking Africa has there been this systematic extension of the donor R and D into development planning and programming in recipient countries. This approach has great value: (1) in the systematic training, recruitment, and career employment of scientific manpower in the donor country, (2) in the continuous evolution of development-oriented research, and (3) in enabling the optimal use of science resources as a motor force for development. On the other hand, such an approach imposes a continuing responsibility that may not be acceptable to the donor, and interlocking relationships between recipient and donor that may not be acceptable to the recipient.

The Values of Science AID

From the viewpoints of both recipient and donor, there are special values in science AID. For the recipient, they include the following:

1. Science AID can create certain of the preconditions for a rapid assimilation of advanced technology, by providing the information infrastructure that modern technology requires, by promoting an indigenous capacity for sophisticated choice and a consequent receptivity to the values of modern technology, and by developing the capacity for adapting advanced technology to local needs and conditions.

2. By advancing the capacity to quantify, accumulate, and systematically organize information relevant to rational choice and practical innovation, science AID enables more effective planning, programming, and control by governments and by private enterprises.

3. A development-oriented science is itself an instrument for attacking environmental barriers to development.

4. Beyond their direct value for economic development, the establishment of science capabilities in the LDC will link its people into the larger world of scientific thought, broadening their cultural horizons, and enlivening their intellectual climate.

From the side of the donor, science AID also has certain advantages:

1. It can conceivably make a larger contribution than any other form of AID to development in relation to the real sacrifice of donor resources.

2. If rationally organized to feed information and experience back into an evolving discipline, science AID will increase the capabilities of the donor country to promote its own technological advance, economic growth, and social well-being.

3. The use of science resources in AID programs can enable donor countries to establish their own research organizations and science capabilities as vital components in the continuing process of world economic growth. This will strengthen the economic and political position of donors beyond the time when the "third world" has industrialized itself and crossed the threshold of high productivity and when, rather than of manufactured goods, it is the export of services, of problem-solving skills and engineering consultancies, of patents, blueprints, and product innovations from the cutting edge of a perpetually advancing science and technology, that will play the key role in international trade.

NOTES AND REFERENCES

1. It is possible to organize the use of the computer as a utility grid, selling its services to small users. This might make it a practicable instrument for decision makers in decentralized market economies. This innovation is yet to be economically devised and practiced.

2. Recent data from *Statistical Abstract of the United States*, 1962, earlier data from *Dael Wolfe, America's Resources of Specialized Talent* (New York: Harper, 1954), p. 31.

3. *National Science Foundation, Scientific and Technical Personnel in American Industry*, pp. 60–62.

Chapter 4

We Can Hire
You Scientist Chaps

AFTER I HAD ADDRESSED AN INTERNATIONAL GROUP OF ECONOMISTS on the capacity of poor societies to assimilate advanced technologies, a black Oxford-trained African dismissed some of my urgings with these words, "We can hire you scientist chaps." Does it suffice that the LDCs hire us scientist chaps, or should they, must they, produce a cadre of their own scientists, their own research institutions, their own *endogenous* science base?

Does Economic Development Require an Endogenous Science?

Granted that science-trained manpower has a significant contribution to make in the process of economic development, need such science-trained manpower be natives or citizens of the country where economic development takes place? Are science-trained Nigerians needed for the economic development of Nigeria? Science-trained Brazilians for Brazil? Science-trained Malaysians for Malaysia? Surely not. It is never, in an absolute sense, necessary that science manpower be indigenous to the developing region or polity. It is always conceivable that science-trained manpower can be brought in from elsewhere to do the job, and, indeed, certainly very many, perhaps even most, of the scientists and science-trained engineers now engaged in development-related tasks come from the rich, high-productivity nations and the technically advanced regions of the world.

A different question might be asked. Are there advantages to a developing society in having its own endogenous science resources? Surely so. Advantages and problems as well. In this chapter some of the advantages and some of the problems of an endogenous science are suggested.

Business Choice and Political Control

Science-trained manpower has a role in the formulation of development policy, in planning and programing economic development, in recognizing the feasibility of implanting particular science-based technologies, in adapting such technologies to the needs and circumstances characteristic of the LDC, and in installing, organizing, and managing those technologies in their practical operations. But the optimal performance of these functions always requires more than simply a knowledge of science. Whether under political directive or through entrepreneural decision, whether in determining feasibility or in modifying technique or in organizing and managing operations, the mastery of science and science-based technology needs to be coupled with an intimate understanding of the social modus operandi, a knowledge of local circumstances, and a sensitivity to prevailing values. It is always advantageous that those who formulate policy, exercise control, or give guidance to those who do, whether in science-related tasks or otherwise: (1) should identify themselves with the goals of a society; (2) should comprehend its complex processes of action taking and choice making and; (3) combine with their expert knowledge, a sensibility to the culture and a familiarity with the social and physical parameters of policy and operations in that society.

There is a presumption that these three conditions are more likely to be met where indigenous manpower is engaged in the tasks of choice and control. Yet it may be that, the structure of privilege, the pattern of class, tribal or regional loyalties, the force of traditionalism, the culture of elitism, habits of subordination, the vested interest in status are such that those in developing societies who are likely to receive the advantages of higher education will be more resistant to innovation, less motivated to promote development, and less trusted by their people or their leadership than are those from the outside.

Development-Oriented Research and Services of the Science Infrastructure

The services of the science infrastructure are necessarily localized and must be integrated into the processes of private and public choice. At least some aspects of development-oriented research must be physically located in developing societies, for these constitute the object of study and the living laboratories of such research. And the information produced by such research, also, to have an impact on development, must gear into development planning and programming. That "gearing in" to public policy or private decisions requires commitment, a knowledge of social organization and of a complex configuration of resistances and proclivities that alway confront novel or decisive action.

This is to say only that if the professional knowledge of the scientist or science-trained engineer is to be translated into action it probably must be coupled with an understanding of physical and social circumstance likely to be acquired only through on-the-spot observation and long personal association. The will to fight through the resistances that block every significant innovation, the skills in social and political maneuver, the subtle and intuitive understanding of the levers of power and the strategies of decision, coupled with the substantive grasp of science-based potentials, are not easily found, rented, purchased, or hired on the world market. They probably must be produced at home.

Foreign scientists may have an enormous contribution to make. Genius and a capacity for dedication are not matters of nativity, and supranational specialization in science is the order of the modern world. Yet the very capacity of foreign scientists to make a link with policy formulation and action taking seems often to depend upon the existence of an endogenous science with whom they can integrate.

The role of the foreign scientist in the developing society is a rather particular one. In some ways it is advantageous. If the scientist comes without a knowledge of cultural proclivities and the established power structure, he comes also without the fears, the sense of inferiority, the expectation of failure that history has inculcated in the minds of the native-born. He thrusts himself crudely and naively against walls and barriers to change, in an effort that the others may have long before abandoned, or never dared attempt. Associated with the "image" of a powerful and technically advanced society, his opinions may carry a weight out of proportion to his private achievement or his personal worth. For the same reason, he may, with equal irrationality, be feared and resented.

Noneconomic Values of Endogenous Science

There are values in establishing an endogenous science aside from any functional contributions to development. It may be a relatively inexpensive way to gain worth in the world's eyes, or to bolster the self-image of a people who for too long have felt themselves inferior. And besides, science is a great adventure. To participate has intrinsic worth. It is a link with the world currents of thought, a channel through which ideas of many sorts may infiltrate, giving vitality to higher education and raising the level of intellectual discourse.

The Economics of Endogenous Science

The capacity for mass communication, an understanding of the machine, mechanical and technical skills, and a capacity to comprehend and to organize complex processes are cognitions that must be built into any society as prerequisites of development. But the scientists and the science-trained engineer can be imported. Or even when it is their own manpower that is being trained to be scientists or science-trained engineers, that training can be given in foreign universities.

Higher education is very costly, and educational institutions of the first quality are hard to establish at any costs. The great universities of Western Europe and the United States, moreover, are heavily subsidized by governments and sometimes by private endowments. It is normally much cheaper for a developing society to educate its science-trained elite abroad than to attempt to duplicate those educational institutions at home.

Another side to the picture is that the university is itself a key institution in social change. It is an incubator of ideas and ideals. It produces the future. Science exists for the sake of the university and its role, as well as the other way around. The university needs pride, competence, independence—if it is to play its role in spearheading change and in producing a social leadership.

Yet, for the most part, universities in developing societies are not leaders but lackeys, imitators and followers, subsidiaries and satellites, dominated by prestigious institutions of higher education elsewhere. Even development-oriented science is the result of the initiative of the great centers in advanced societies.

Science, in some of its aspects, is universal; it is precisely those universal aspects, those upper reaches of the pure and fundamental, that are almost entirely irrelevant to economic or political choice, or to the specific problems and technical needs of development. Yet it is to these upper reaches that the brilliant student from the developing economy tries to climb when he comes to the university in one of the world's great centers of learning; and there he seeks to remain even after he returns to his native land. This is for a number of reasons:

1. In Academe generally, and perhaps particularly in the European universities, pure and fundamental science is glorified.

2. The very universality of pure and fundamental science has a peculiar economic value to the student from the developing country, for it offers him the possibility of escaping from the onerous condition of his own society to a more lucrative and perhaps a more exciting career elsewhere. And indeed, in so doing, he may make a considerable contribution to the world science, though his work will have no relevance to the economic development of his own country.

3. The applied science taught in the colleges and universities of advanced countries is likely to be of minimal relevance to the needs and problems of the developing economy. Before it can be taught, a development-oriented science or a science-based technology geared to the needs of the developing economy must first have evolved. This requires research, experiment, application, implementation, continuously directed toward the real problems of developing economies, with results and experience fed back into an organized discipline. Even where such knowledge has evolved, it is not likely to have found its way into the university curriculum in Western Europe or the United States. What is taught there will, naturally, be suited to conditions and interests of a Western clientele.

4. In spite of the academic bias of his education, the young science graduate looking for work in the United States and Western Europe will most likely be drawn into industry or government, and be trained there, on the job, to the skills of science application. But the young science graduate who returns to the traditional culture and the craft economy of his birth, will find no managerial or R and D opportunities to reshape his skills and to refocus his efforts. His peers may look upon him as an

oddity or as an object of veneration—either way he will not be considered as a resource to be rationally deployed for practical purposes. Uncorrupted by the lures of Moloch (since there are no such lures to corrupt), he remains dedicated to the "fundamentals." When his government provides him with research facilities as an act of faith and reverence, he, following the path of his teachers, will engage in research detached from the mundane world of industry and trade.

On this account, one finds in these traditional societies the extraordinary phenomenon of high-caliber science-manpower without any practical outlet for their energies and skills, existing as an unexploited and wasting resource, or engaged in work that cannot conceivably contribute to economic development. In these terms one can easily understand the Chinese moratorium on a pure and fundamental science.

Chapter 5

Science-Based AID in France

The Record

FRANCE'S RECORD AS AN AID DONOR IS TRULY A FORMIDABLE ONE
—in relation to its resources, the most impressive in the world. Writing in
1963 it was then possible to say:

> In its sheer magnitude, French technical assistance is not comparable to
> that of the United States. But, in terms of French resources, the French
> commitment to support economic development is unparalleled and (as will
> be shown subsequently), the French organization of its science resource, as
> a part of its technical assistance program, is a model for the entire world.
> The amount of aid given per capita in France is the highest in the world.
> The amount of aid, given as a proportion of the gross national product in
> France, is the highest in the world. Thus, French public aid for economic
> development is *more than double* the proportion of public aid in relation to
> GNP in the Federal Republic of Germany, the United States, the United
> Kingdom, the Netherlands, or Japan. When private investment and gifts are
> also included as "aid," France remains far in the forefront.

France is still in the forefront, but its relative position is no longer what
it was. In terms of the net flow of resources, France is now surpassed by
the Netherlands. And while AID has increased elsewhere, in France it has
in real terms declined. In 1969 public AID as a proportion of France's gross
national product was less than half what it had been in 1961.

French AID is predominantly bilateral, though there has been a slight
trend toward multilateral arrangements, i.e., multilateral AID as a proportion of total public AID increased from eight percent in 1962 to eleven
percent in 1970,[1] reflecting perhaps the loosening of ties between France and
her former territories.

France concentrates AID in its territories and former territories, particularly in the French-speaking, or Francophone, parts of black Africa. There has, nevertheless, been something of a trend toward diffusing French AID more widely, outside the French Zone.

Table 5
DISTRIBUTION OF FRENCH AID

| | As Percent of Total | | | Change between 1964–69 | |
	1967	*1968*	*1969*	*(millions of francs)*	*(%)*
North Africa	27.0	23.5	25.4	+ 49.3	+ 15.0
African States and Madagascar	60.7	57.0	57.1	+ 114.5	+ 15.6
Rest of World	12.4	19.6	17.5	+ 108.6	+ 72.0

SOURCE: "Gosse Report."

According to the earlier "Jeanneny Report,"[2] this concentration raised the per capita level of assistance in the French Zone far above what it was anywhere else in the developing world. Thus, in 1961 AID per inhabitant of the French Zone was in the neighborhood of $18.50, whereas its average elsewhere was about $6.00 per head.[3]

> If one compares more precisely the aid given to countries which receive the greatest help from the industrial nations, the differential advantage of the territories of the French Zone is even more marked. Thus, in 1961, the French overseas territories received aid to the extent of $100 per inhabitant, and Algeria received nearly $35 per inhabitant; while South Korea, which is the country outside of the French Zone which has most benefited from aid programs, received only about $10 per inhabitant.[4]

The special AID advantages accruing to inhabitants of the French Zone, while not as great as formerly, are still considerable, compared to those elsewhere in the developing world save perhaps for those unfortunates caught in the margins of Great Power conflict, the Cold War Zone, where nominal AID shades into the support of militarism, the machinations of imperialism, and internal suppression.

More than any other donor, the French have emphasized technical assistance rather than financial supports. Thus, in 1969 "technical cooperation" constituted 52 percent of total French public AID, as compared to 31.3

Table 6
PER CAPITA AID RECEIPTS IN SELECTED COUNTRIES
Net Official Receipts

Countries	% of Imports	% of GNP	$ Per Capita
FRENCH ZONE			
Algeria	16.40	3.11	8.10
Tunisia	38.35	9.55	21.48
Senegal	26.44	5.08	11.41
OTHERS			
India	39.45	2.42	2.03
Indonesia	36.82	2.76	2.66
Thailand	6.90	. .	2.19
Ghana	22.73	3.64	8.66
Peru	4.99	1.06	4.01
Ethiopia	22.66	2.69	1.72
Nigeria	12.12	1.69	1.62
COLD WAR ZONE			
Pakistan	35.49	2.42	3.74
South Korea	29.58	5.49	10.33
South Vietnam	52.64	14.86	26.02
Turkey	27.25	1.87	6.48
Taiwan	8.12	1.71	5.33
Laos	216.38	33.21	23.74
Dominican Republic	23.43	4.55	13.20
Greece	3.36	0.58	4.95

SOURCE: Data is from the OECD, *Development Assistance 1970 and Recent Trends,* p. 35, "countries to which the annual net flow of resources under DAC official bilateral programmes and from mutilateral agencies exceeded 140 million on average in 1967–1969: selected statistics."

percent for Germany, 29.5 percent for Great Britain, and 22.5 percent for the United States, or to the average of 26.9 percent for all DAC countries.[5] In France, moreover, as elsewhere, the trend has been toward a greater emphasis on technical cooperation, e.g., from 40.5 percent in 1964 to 52 percent of total public AID in 1969.[6]

So much for the record. But the records needs to be explained. Why has France been willing to sacrifice so much larger a proportion of her resources to the purposes of AID than other advanced nations? And why has France placed so much greater emphasis on technical cooperation than the others?

Raison d'Être

French AID is, in effect, the French effort to promote economic development in former French territories and in Francophone Africa. As such, it could not but partake in France's effort, and the French approach to the rebuilding of its own economy and the promotion of economic growth in France—an effort, incidently, that has, by the record of growth rates, been extraordinarily successful in recent decades. The French, far more than any other Western people, have tried to organize and rationalize the growth processes under a national plan, "modernizing" technology and attempting to promote technological advance through public investment in science and industrially oriented research.

This form of economic and industrial planning in France (and in the effort to promote development outside of France) reflects the particular outlook and competence of French political elites, characteristically the highly science-trained polytechnician.

Technical cooperation, as a part of French AID, was swollen by another factor quite unrelated to any science-based planning of development. The French offered qualified university graduates the option of working, presumably as teachers, in developing countries, in lieu of their military service. Thus, most French technical assistance consisted of young teachers, enthusiastic and able no doubt but hardly qualified to promote industrial development in the LDCs where they worked, and reinforcing an educational system that followed the classical French model, and, indeed had its apex in the French university—a system that functioned to select and train an elite leadership rather than gearing itself to inculcate the basic cognitional requisites of development into the whole society.

Table 7 shows the distribution of technical personnel, as part of French technical cooperation.

If this explains something of the organization of French AID, it hardly accounts for its magnitude or its concentration in black Africa. The magnitude of French technical assistance is certainly not to be explained as an effort to stave off communism nor by any role which France supposes itself to be playing in the Cold War. Nor has its unusually large AID program been based on any rational expectation that it will yield France any significant commercial advantages. Nor has there been any attempt to so justify

Table 7
DISTRIBUTION OF TECHNICAL PERSONNEL IN FRENCH
OVERSEAS TECHNICAL COOPERATION—1970

Country	Teaching (%)	Other (%)
Near and Middle East	87.4	12.6
Indochina	85.0	15.0
North Africa	84.4	15.6
Africa (Non-Francophone)	82.0	18.0
Latin America	77.8	22.2
Elsewhere in Asia	76.5	23.5
Francophone Black Africa and Madagascar	66.0	34.0
TOTAL	77.0	23.0

SOURCE: "Gosse Report."

AID expenditures before the French public. This was made explicit by the "Jeanneny Report" in 1963, which, speaking to the French public, insisted that in spite of French AID there had been a decline in the French share of trade with the countries of the French Zone, that no monetary or commercial advantage had been accrued to France as a consequence of its technical assistance program, and that none could be expected in the future.[7] What then motivated France for a time at least to far exceed the AID effort of all others?

The "Jeanneny Report" found the essential motivation in "le besoin de rayonnement"—literally the need to shine, to send out the rays of an "esprit" to reanimate "la gloire" of "la civilisation Française." This need of France to reassert herself as a center of world culture and thought, and in this age of science and technology as a world center of technology and science, was not difficult for an American immersed in the French circumstances to understand.

France has emerged from the depths of the past. She is a living entity. She responds to the call of the centuries. Yet she remains herself through time. Her boundaries may alter, but not the contours, the climate, the rivers, and the sea that are her eternal imprint. Her land is inhabited by people who, in the course of history, have undergone the most diverse experience, but whom destiny and circumstances, exploited by politics, have unceasingly molded into a single nation. By reason of its geography, of the genius of the races who compose it, and of its position in relation to its neighbors, it has taken on an enduring character that makes each generation of Frenchmen dependent on their forefathers and pledged to their descendents. Unless it falls apart, therefore, this amalgam, on this territory, at the heart of the world, comprises a past, a present and a future that are indissoluble. Thus, the state, which is answerable for France, is in charge, at one and the same time, of yesterday's heritage, today's interests, and tomorrow's hopes.[8]

Consider the great cultural continuities, moving as streams through history, each embodying a cast of thought, a complex of languages and emotional proclivities and manners and traditions, including as subcurrents many nations, diverse configurations of values and experiences and organizations. None are "pure," yet each has its character. And no sensitive man who moves cross-current from one to the other will fail to sense (though no one can fully grasp) their differences. Within the European stream three great currents are distinguishable: the Germanic, including the British, North Americans, Germans, Scandinavians; the Slavic, clustering in Eastern Europe; and the Latin, including the South Americans, Spanish, Italians, and French. In recent decades the United States has become the power center of the Germanic world. The Russians are pre-eminent among the Slavs. Because of their economic achievement and the dynamism of their technology, the voice and the outlook of the Germanic and the Slavic cultures have become dominant. Whereas the Latins, whose culture is the most ancient and the most deeply rooted in civilized history, have been increasingly dominated, following behind in the paths that the others open. Of all the Latin countries, France alone stands at the industrial, technical, scientific, and cultural forefront.

With reason, France can feel herself in danger, not of being conquered, destroyed, or subjected to alien rule, but of wavering as a bearer of history, of its voice being drowned and lost in the din of greater numbers suffocated by the output of a more productive technology, of being isolated and passed by, of losing its audience and its own sense of mission, of losing the belief in its particular qualities and unique worth. That danger for France is a danger as well to the Latin world. For France, at once creative and disciplined, is perhaps the only Latin nation at this point in history able to carry forward and to express that great and ancient cultural continuum. Without France, not as a world power but as a world voice, as a world nexus of thought and culture, the Latin world might become a static and stagnant backwater, uncertain of its identity and direction, unable to make its distinctive contribution, as, for so long, the Chinese, the Indian, the Arab worlds have been. For this reason, French AID, as a reaching out and as a means of maintaining itself as a distinctive force and as a voice that is heard, has acquired a special force.

One may agree or disagree with the above as an interpretation of history and of social crisis. Our intention, however, is not so much to interpret history as to articulate one outlook that underlays French policy; thus, to explain the magnitude of French AID as a means of holding Francophone Africa within the realm of the Latin "esprit et civilisation."

And, if indeed, this is the premise of French policy, one wonders whether France does not err in pouring so great a part of her resources into the Stone Age backwardness of Equatorial Africa and into the hostile environment of North Africa, rather than into the Latin societies of eastern and southern Europe and South America, and among other civilizations with whom a rapport might come easily.

Where does France belong? Where should she seek to find her base and her place and her purpose: in Europe as a part of an integrated Europe? among the French-speaking peoples of Asia or Africa? or in the great community of Latin cultures?

Geography fixes her in Europe, and considerations of economic growth and security demand European integration. But all the inclinations of France's northern neighbors (Scandinavia, Great Britain, or a healthy Germany that hopefully at long last has escaped the dark cave of tribalism) draw them into another community of culture and cognition that merges with and is now centered in the United States. In that community of peoples, France is perpetually alien.

Conquest and colonization, accidents of nineteenth century imperialism, that did implant among the elite at least the language and outlook of France, complete with the extraordinary loyalty to the "idea" of France shown by black Africans during World War II, tie France to Equatorial Africa. But this is a tie to a hard environment and a poor land, thinly populated, that has only just crossed the threshold into civilization.

Another quality of the French *esprit* accounts for this African attachment. Power and property are gone, but deep emotional roots into the land remain, for the French are a peasant race with a peasant's parsimony of and devotion to the soil and the earth itself.

> I have read with great interest your interpretation of our French technical assistance policy in France. What you say seems entirely true. But, in my opinion, there is another motivation that attaches us, for better or worse, to our former possessions. The French in their roots and origin are a peasant stock, a people profoundly attached to the land even when the land is poor. The officials or the scientists who have worked under colonial regimes felt themselves proprietors of a wretched earth. They had to make that land viable as a matter of pride (the proprietor must be proud of his property) and as an obligation to the land itself, whether or not their efforts and investment were economic. Thus, we tried, without much success I admit, to improve the impoverished earth of Africa as though we were peasants working our own fields in France. After the independence that way of looking at things has remained.[9]

It is surely arguable that France's *besoin de rayonnement* might better be served by a commitment to raise the technological capabilities, the economic powers, and the cultural force of the Latin world, for which France is a natural center—that France is the society best able to give coherence and force to the outlook and to find the way to realize the aspirations of the numerous and immensely vital Latin people.

In this light, the "Jeanneny Report" urged in 1963 the diffusion of French AID beyond its traditional confines; but a decade later, reviewing such "outside" AID, the Gosse Commission asked for a withdrawal and a refocusing of efforts within the French Zone, arguing that France should concentrate its resources where they are most needed and where they can be the most effective. On both counts, it finds black Africa to deserve the highest priority. There the poverty is greatest, and there also France operates from a sufficiently dominant position as to be able to integrate its AID effectively into (and indeed to shape) general development programming and planning. According to the "Gosse Report," France might also leave its mark through AID in the Mideast, and exploit the high potential payoff of increased cooperation with Anglophone Africa. But in Latin America— where French AID is a small and insignificant element, and where France must operate amid a confusing diversity of doctrines and policies, where the French contribution is necessarily marginal to that of the United States—

the Gosse Commission advised withdrawal. Seen from the standpoint of the Gosse Commission, AID is not the instrument through which France might generate, integrate, cohere, and express the creative and constructive powers of the Latin world.

In any case, the barrenness and emptiness of Equatorial Africa has given a peculiar shape to the organization of French AID and has given France a certain advantage in its development planning and programming task, particularly for the rational deployment of science resources. Starting almost with a clean slate, AID-supported development could be organized step by step, from start to finish, as a structural and operational whole. For societies such as India or Brazil—where the approach is to the complex problems of a highly evolved social and economic structure—quite another strategy is required.

Limitation in Science Resources

French academic institutions, rooted in medieval tradition and the Napoleonic establishment, developed as an instrument for selecting and training a ruling elite of pedagogues and bureaucrats to be fitted precisely into the niches of the social and political hierarchy. During the postwar decades, as French society opened itself to the force of mass aspirations, the ancient educational edifice could not bear the increased educational burdens placed upon it, nor satisfy the expectation of new generations. It collapsed, and the cumulative student frustration exploded in the extraordinary Revolution of May 1968. Since then, the system of education has been in a state of radical transformation.

Traditional French science has been pure, predominantly academic, based on the efforts of a few gifted individuals. The team effort, long-range and systematic, geared to practical payoffs in the economy is a new departure. The deliberated use of science in support of economic growth is a basic innovation and the science resources available for that effort are relatively meager. In important research areas, science manpower is thinly spread. Nor can this lack be wholly compensated by the vigor of youth. This must be taken into account in evaluating the French organization of science resources in AID programs, which is our particular concern.

Amid a welter of frustrations and discontents, French science is growing, replenishing itself, increasing in number and in research potency. The magnitude of the French effort to replenish and expand its capabilities in science, is suggested by the educational data. Thus, taking 1958–59 as the base line, the numbers of entrants into universities had, by 1966–67 increased in France by 131 percent, in the Netherlands by 110 percent, in the United Kingdom by 89 percent, in the United States by 64 percent, in West Germany by 36 percent.[10] A comparison of university entrants in these countries is shown in Table 8.

Table 8

UNIVERSITY ENTRANTS IN SELECTED COUNTRIES

Country	1962–63	1966–67
France	88,340	133,881
West Germany	52,049	63,738
United Kingdom	34,647	54,956
Netherlands	8,425	13,304
Japan	201,125	292,958
United States	776,852	995,000

SOURCE: OECD, *Development of Higher Education* (Paris: 1970).

The Organization of French AID

The control functions for French AID are quite widely dispersed and to list the responsible departments tells little as to where initiatives originate, how policy is made, or who controls and coordinates whom. The Ministry of Foreign Affairs and the Ministry of Economics and Finance bear normal responsibility for multilateral AID. The Ministry of Foreign Affairs, the Secretary of State for Foreign Affairs, the Ministry of Economics and Finance, the Ministry of National Education share responsibility for bilateral AID. Moreover, since development-oriented science in France is integrally related to and is functionally inseparable from other research and science-based activities, the ministries concerned with agriculture, forestry, fisheries, education, and the general organization of science are involved in organizing science resources in support of technical assistance also.

Below and apart from the ministries, to which they are variously answerable, are a number of organizations with research (or other science-based) capabilities related to developmental needs, supported wholly or in part by government funds and involved in technical assistance. These are the operational entities for science AID. They are substantially autonomous in administrative matters, and almost wholly autonomous in their professional activities. They may: (1) systematically supply information useful for developmental planning, (2) develop or support the operation of the science infrastructure, and (3) participate in development planning, developmental projects, and technical innovation.

Nominally, research programs or research-based development programs in the French Zone are proposed by the political authorities of recipient governments or territories, and are negotiated, normally on a cost-sharing basis, with the appropriate French ministries. Agreed-upon programs are then forwarded, as ministerial instructions, to the autonomous science organizations. In fact, the initiative has usually come from the science organization itself, which has worked out an action program with cognizant entities in the LDC. These programs then move up through political and diplomatic channels to the level of intergovernmental negotiation and down again as ministerial instructions.

Development-related, science-based organizations do not operate in isolation. They are set in a matrix of other French development-oriented organizations designed to integrate science information and initiatives into working projects, to harmonize projects and general planning, to supply the loans and credits required to underwrite the transformation of technology, and to arrange for the marketing of new outputs.

French development-oriented research organizations have also educational functions. In some instances, the educational objective is primary. In building their scientific and technical staff, they actively recruit Africans as field technicians and, through French institutions of higher education, train them for research posts. French laboratories, field offices, and AID agencies are open to foreign scientists, and French research organizations send their young recruits abroad to work and train in science institutes, considered as outstanding in particular research areas throughout the Western World.

Thus, development-oriented research is systematically integrated into development planning and programming in the French Zone. In effect, a sector of the French R and D cum economic-developmental establishment parallel to that which operates in France has been set apart in support of technical innovation and economic growth in the territories and activities of the French Zone. Scientific organizations operate under the bilateral surveillance of the French and of the recipient governments. Regardless of who pays the bill, the science organizations are obliged as a prerequisite to the continuation of their activities, indeed of their organizational existence, to serve and to satisfy authorities in the recipient countries as well as in the French government.

In its technical assistance to countries outside of the French Zone, the French approach is pragmatic, flexible, and unsystematic.

Characteristics of the Development-Oriented Research Institution

Typically, the development-oriented research organization is supported by the French government, with some of its income derived from work done under contract with foreign governments and with international agencies.

It is largely autonomous with respect to budgetary administration, recruitment, and the planning and programming of its professional activities. Such an organization ordinarily operates under a Conseil d'Administration (Board of Directors). Its high officials are sometimes civil servants, seconded by the appropriate ministry. Its Board of Directors include representatives from: (1) the ministries concerned, (2) sometimes commercial interests, (3) sometimes the governments of AID recipients, (4) political and scientific "personalities," and (5) scientists and research administrators from related centers or educational institutions. There are also informal channels of communication and coordination through the intermediary of the university, where upper-echelon scientists generally hold professorial appointments. Moreover, the French intellectual establishment is still comparatively homogeneous, small and tightly knit.

French development-oriented research organizations have their headquarters in Paris, and carry on the bulk of their experimental research in the French Zone of Africa. Besides its administrative and policy-making functions, the Parisian center provides back-up services to research in the field and performs certain research tasks requiring highly complex equipment and very specialized skills. The Paris center supplies the documentation, library facilities, information services, publication outlets; it arranges for conferences, seminars, and symposia, for student-exchanges, which, taken together, keep its research staff, whether in Paris or in the field, in touch with world science. The ultimate location of the research activity · itself will depend on counterbalancing attractions: need for expensive specialized equipment and direct contact with a range of science specializations draws research toward the Paris center, and the need to explore or to experiment upon the characteristic physical or social environment, to follow through on results, and then have a feedback from applications draws research toward the field.

Development-oriented R and D organizations in France are very numerous. Thus, in 1963 l'Institut d'Étude du Développement Économique et Social, in their series "Tiers Monde" published the monograph *Recherches et Applications Techniques et Matière de Développement Économique et Social et Répertoire d'Organismes Français,* which lists some two hundred and twenty-five *development-oriented* science organizations actually engaged in R and D or in related training activities, geared to technology and policy in LDCs—including, for example, twenty-nine different research organizations specializing in diverse problems of agriculture for developing economies; two in general science in relation to development; nine on preventive medicine, human health problems or nutrition in developing countries; eight on hydrology, sea water conversion or water resource surveys in those countries; one in metal fabrication research; three in electrification studies; one in research in textile production; eight in fields of chemical industries or chemical research; one in packaging research; two in studies pertaining to radio and television communication industries; one in research on roads and bridge building; one in research on fisheries; three in studies of railroad organization; two in iron and steel research; one in solar energy research; one in atomic energy research; two in research related to the ceramic industries, one in electronics research; three in research in telecommunications; nine in studies of building and construction; one in low temperature physics, one in geophysics; two in research on natural gas problems; three in geology and minerals; one in forestry studies—all focused on the problems of and development-planning for low-productivity economies.

Our particular concern will be with those science-based institutes that are central to government AID programs, including those focused on: (1) tropical export crops, (2) subsistence agriculture, and (3) general science in relation to development programming and planning.

The first development-oriented R and D thrust in France was intended to support tropical plantation agriculture, producing crops exported to

France of interest to consumers and manufacturers in France and drawing upon the experience of the administrators and technical experts who had formerly serviced French plantation owners in colonial territories. After World War II, institutes incarnating the old expertise were linked to the destiny of the new African states, and to the efforts of those states to promote exports as a source of foreign exchange. Starting with the I.R.H.O. in 1942, five institutes were established in the following order:

(1) The "Institut de Recherche pour les Huiles et Oléagineaux (I.R.H.O.) (Research Institute for Oils and Oil Seeds);

(2) The "Institut Français de Recherches Fruitiéres Outre-Mer" (I.F-.A.C.); (French Oversearch Fruit Research Institute);

(3) The "Institut de Recherches du Coton et des Textiles Exotiques" (I.R.C.T.E.) (Cotton and Exotic Textile Research Institute);

(4) The "Institut de Recherches sur le Caoutchouc en Afrique" (I.R.C.A.) (African Rubber Research Institute); and

(5) The "Institut Français du Café et du Cacao et autres Plants Stimulantes" (I.F.C.C.) (French Institute for Coffee, Cocoa, and other Stimulating Plants).

As an example of these, the activities and achievements of the I.R.H.O. will be examined in some detail.

Institut de Recherches pour les Huiles et Oléagineaux

In its *Bilan des Resultats Acquis,*[11] this institute for research on oils and oleaginous products, reported on its achievements:

> At the end of World War II the I.R.H.O. started in black Africa an action that combined research and application for the purpose of developing the production of natural oil products. Thanks to the results it obtained in seed and plant selection, in the development of fertilizer practices, in the struggle against pests and plant disease and in the development of agricultural technologies, it has enabled the tenfold increase in palm oil yields, the quadrupling of copra yields, the doubling of peanut yields. Different countries have carried through development plans based on its proven techniques.[12]

The I.R.H.O. is a substantial operation, employing more than 2,500 people, 160 of whom are science-trained. It operates on five continents, maintaining permanent experimental or production operations in twenty countries with temporary programs in a dozen others.

It is organized in the form of an autonomous corporation, supported by the French government, entering independently into contracts with other governments and international agencies. It maintains connections with the French ministries, African governments, and trade circles. Its board of directors includes representatives of the French ministries, of aid-recipient African governments, and others who, as individuals, bring to bear a knowledge of related science or of the interests of the trade. These multiple links are reflected in its handsome publication *Oléagineaux Revue International des Crops Gras,* which combines the characteristics of a scientific journal and a trade review.

The bulk of its activities are carried on in experimental stations, test plots, and operational programs in tropical areas. Paris headquarters is not only a center of administration and liaison, but also of research, documentation and technical services. The Institute has a world reputation for excellence, reflected in this letter from the vice president of the Rockefeller Foundation, after a world survey of research in agriculture:

> We were particularly impressed with the research effort on oil palm—the one international program which we were able to learn about in some detail. I think it is safe for me to say that I have never before seen such a comprehensive, sustained, and imaginative single research effort on any crop, with the possible exception of the work on sugar cane by the experiment station of the Hawaiian Sugar Planters' Association headquartered in Honolulu. We wish to congratulate you especially on the imaginative use of genetics, coupled with advances in other sciences, in your constant search for quantum jumps in productivity and profitability with this crop. When such singleness of purpose and dedication to achieving it are combined over such a long period of time as has been maintained by IRHO, maximum success is assured.[13]

One of the I.R.H.O. personnel has been invited to head an international institute for research in agriculture in coconut products that the Rockefeller Foundation now plans to establish in the Philippines.

The R and D activities of the I.R.H.O. are of the following sorts:

(1) *The control of plant diseases and pests.* The I.R.H.O. has established norms and criteria for the surveillance of plantations, and has developed the means and the system of control against the principal palm predators. The means for the effective prevention and control of major diseases attacking

the palm tree have also been developed. In evaluating and selecting materials and equipment, the Institute works directly with manufacturers to develop and distribute products best adapted to the circumstances of intended use. Similarly, for coco trees and for peanuts. For the latter, research efforts have concentrated on the treatment of seeds before and during storage, and on the elimination of contaminated products.

(2) *The development through selective cross germination of more productive hybrid varieties, of disease-resistant strains, and of varieties adapted to particular climatic conditions.* Experimentation by the I.R.H.O. in the selective germination of palm oil varieties has been of an unprecedented scale. The Institute supplies from five to eight million seeds annually to service the cultivation of from fifteen to twenty-two thousand hectares (a hectare equals 2.47 acres).

In the case of the coco tree, artificial selection and crossbreeding is a more recent development. A tract of five hundred hectares has been set aside for the cultivation and study of diverse varieties, constituting now the richest gene stock in the world. Hybrids have been developed that fully mature in seven years, then producing approximately four tons of copra per year as compared to local West African varieties that produce two tons of copra annually when they reach maturity in their twelfth year. For coco trees, artificial germination is not feasible, and using its garden seeds for natural germination, the I.R.H.O. is the only organization in the world now able to supply selected seeds on a large scale. Twelve hundred hectares had been planted in its seeds by 1970, with fifteen thousand hectares planned by 1975.

The I.R.H.O. developed a computer-based mathematical technique *(methode Larroque)* for the analysis and seed selection for peanuts that enabled superior varieties to be developed and disseminated within four years, for the rapid amelioration of peanut culture in the Congo, Casamence, and the Upper Volta. Using classical methods, a late-blooming variety of peanut resistant to the rosette virus that, with a slight fertilization, is able to increase yields by five to ten times has now been developed. Similarly, varieties have been produced adapted to conditions of heavy rains and to semiaridity, and with growing seasons of diverse lengths suited to the different regions of the African continent.

(3) *The development of land cultivation, processing, and product control techniques.* In 1960 the I.R.H.O. inaugurated a system for the mechanical

preparation of forest soils, which by 1970 had been used to bring more than fifty thousand hectares of palm oil trees under cultivation, often under circumstances where plantings might otherwise have been impossible. Similarly, mechanical techniques of soil preparation have been applied to savannah soils.

By protecting seedlings in plastic sacks, plantings required per hectare have been reduced by twenty percent, and the time required to obtain a plant has been reduced by nearly a third. Techniques of early castration have reduced the time span required to bring the trees into production and stabilized their yields during dry spells. The I.R.H.O. has developed a range of techniques to conserve irrigation water for use in areas of excessive dryness, as well as general techniques and criteria for the control and conservation of fertilizer elements and chemical inputs.

The I.R.H.O. has constructed and manages eleven palm oil factories. It has cooperated in the design and organization of twelve others. It maintains a direct relationship with the manufacturers of machines and equipment, promoting the in-transfer of advanced technology, leading to automatization of oil-producing operations.

Similarly with coco trees, technology has been developed for all stages of cultivation and production, including deforestation and land preparation (of great importance for disease control in this instance), for protection of the soil against erosion, for precise fertilization control, for the protection of seedlings. With the shortage of manpower as a bottleneck to production expansion, it has, in cooperation with equipment manufacturers, developed and introduced the mechanical shelling of the nut, the extraction and drying of the copra, the preparation of the fresh coconut for export. A technology for peanuts, adapted to the needs of small holders and peasants has been developed and disseminated.

(4) *The development of product and by-product outlets.* Product and by-products outlets have been developed in the manufacture of hard board from palm stalks and groundnut shells. New fatty acid derivatives from methyl esters and from glycerides of palm oil, copra oil, and shea butter have been prepared for use as plasticizers, fungicides, insecticides, and detergents.

(5) *The modification and adaptation of agricultural theories and practices evolved in temperate zones to the general conditions of the tropical environ-*

ment and the special needs of oil crops. An interesting early example were the findings made in respect to the use of mineral fertilizers on tropical soils. The nutriment needs of temperate and tropical soils have been found to be, in one sense, opposite. Temperate soils must be continually fertilized with organic materials. Soils in temperate zones sometimes also need mineral fertilizers, but from year to year the earth in temperate zones retains a high proportion of the mineral fertilizer which it has received so that, once the soil has been saturated with a particular mineral fertilizer, it tends to require only marginal renewals in subsequent years. In tropical zones the rapid spontaneous production of organic matter virtually eliminates the need for organic fertilizer. But, for a variety of reasons (e.g., the soil structure, the concentrated rainfall, the rate at which minerals are consumed in the spontaneous growth of organic matter), the mineral content of tropical soils is very rapidly depleted. Hence, mineral fertilizer must be massively and continuously introduced to raise and to maintain the fertility of tropical soils. The I.R.H.O. has been a leader in the study of the comparative fertilizer requirements of tropical soils and in propagating the principle of the massive and continuous application of mineral fertilizers to such soils.

The I.R.H.O. has: (1) itself constituted as an important component of the science infrastructure, and (2) in some instances, spearheaded the planning and implementation of development programs. These can be illustrated by two specific instances.

The I.R.H.O. has developed a quite extraordinary service to guide fertilizer practice in its agricultural territories. Working through the experiment stations and field agents, thousands of carefully marked leaf samples from oil-producing plants are continuously forwarded to Paris. Each is submitted there to a foliar analysis from which the nutritional lacks or imbalances of the plant can be deduced. The results are recorded according to the areas from which the leaf samples were taken. Through a computer-based statistical analysis the nutriment-conditions of soils from which the leaf samples have been taken is quickly determined. On the basis of this knowledge, the soil's nutriment needs are diagnosed and an appropriate fertilizer formula is prescribed for each area and immediately forwarded to the field as a guide to practice. By 1970 more than forty thousand foliar analyses had been made as a guide to plant nutrition on more than fifty thousand hectares planted in palms. These data serve also as an experimental base for determining the relationship between nutritional inputs and plant productivity.

The Ivory Coast government, wishing to diversify away from a high concentration in coffee and cocoa, asked the I.R.H.O. to study the possibilities of expanding the cultivation of oil palms. The I.R.H.O., which had experimented extensively in the area, submitted a plan in 1961. The following year it was asked to put that plan into operation. In the first stage (1962–65) (financed by the "Fonds Européen"), two oil factories were built and 5,000 hectares of plantation were brought under cultivation. By 1970, 67,563 hectares were under cultivation, producing 42,000 tons of palm oil annually.[14]

The plan was to build modern oil-producing factories, and simultaneously to bring oil palm plantations into production in close physical proximity to each factory in order to provide that factory with the supplies it requires for profitable operations. The factories and plantation blocks sufficient to insure that each factory has at least sufficient oil-nut supplies for break-even operation would be owned and operated by an Ivory Coast government enterprise, Sodepalm. It was envisioned, however, that the bulk of the production would come from the private holdings of peasant agriculture. The particularly difficult and delicate problem would be to induce the peasantry to convert their land into the production of oil palms, where no yields could be expected for five years, and to train them in the modern techniques of oil palm cultivation. For this purpose, a complex scheme of demonstrations, training, conditional loans, and periodic inspections was devised. In creating the new industry, the I.R.H.O. spearheaded the developmental process.

The record suggests solid achievement and unresolved problems as well, particularly in integrating the peasant palm culture into the needs of a factory system for a predictable and controlled flow of inputs. By the end of 1970, out of an Ivory Coast total of 67,563 hectares, Sodepalm and Palmivoire had planted 54,254 hectares, of which 36,982 hectares was in industrial plantations, and 17,272 hectares in village plantations. In that year Sodepalm-Palmivoire produced and sold 20,602 tons of oil, for the first time entering the export market. Although it had planted eighty percent of the total Ivory Coast hectarage, Sodepalm and Palmivoire produced slightly less than sixty percent of the commercial output of palm oil. Evidently a great part of that lag resulted from difficulties encountered in organizing the harvesting and collection from village plantations. On that account, the Ivory Coast government declared February 1970 "The Month of the Palm Tree," and undertook, with a high-pressure national campaign, to promote harvesting efforts, with numerous arrangements made to facilitate and to motivate more effective harvesting and collection in the villages.

Policy for Plantation-Oriented Research

It is hardly possible for the outsider, particularly one who is unable to visit the research centers and areas of application throughout the world, to evaluate the different French R and D institutes organized to promote tropical export crops. Not all, it would seem, have attained the pre-eminence of the I.R.H.O. Others have been established more recently and they operate in quite difference circumstances. Thus, the Institut Français du Café et du Cacao (I.F.C.C.), which attempts to develop the cultivation and production of coffee, tea, cocoa, and cola in tropical countries, for example, was established in 1958,[15] and perhaps a longer time should be allowed before major results in plantation development and in agricultural practice can be expected. Yet the I.F.C.C. is also now a flourishing, highly professional organization, whose services are in demand throughout the world. Indeed, two-thirds of its financing is from outside of France. It employs approximately a hundred researchers and fifty technicians, aside from those engaged in administration, documentation, and field operations. Its research centers are located in France, the Ivory Coast, Cameroun, the Central African Republic, and Madagascar, and it also has twenty-six permanent experimental stations or pre-extension centers elsewhere. It has not yet evidently achieved the tie-in to developmental planning that distinguishes the I.R.H.O. It emphasizes research and the dissemination of scientific information[16] and claims leadership in the developing the means for dealing with the red rust that attacks the coffee plant.

Or, to take another example, the Institut de Recherches du Coton et des Textiles Exotiques[17] (I.R.C.T.E.) concerned with cotton and fiber-yielding plants for sacking and cordage, such as the hibiscus, sisal, jute, ramie, Urena. Beyond the point of fiber selection and ginning, another R and D agency carries the responsibility for the development of textile technology. The I.R.C.T.E. was formed in 1946. In 1971 it employed seventy-four researchers, operated major research centers in the Ivory Coast, Chad, and

Madagascar, and had fifteen other regional centers and experimental stations. It did not fill a research and development vacuum as the I.R.H.O. did, but entered a field dominated by cotton-oriented R and D in the United States. American R and D, however, is not geared to promoting and ameliorating cotton culture among the LDCs. Thus, for example, American research has produced varieties that are without the gossypol gland and hence are free from the toxic substance produced by those glands. But it is the I.R.C.T.E. that concerns itself with the food potential of seeds produced with this new variety, undertaking to develop technology appropriate to utilizing the high protein cotton seed as a cereal food for the people in Africa.

Compared with the I.R.H.O., moreover, the I.R.C.T.E. does not enjoy the relative ease in the dissemination of techniques that would be the case if its dealings were with large plantations, for in Francophone Africa cotton is mostly raised on small plots by peasant farmers.

> The culture of textile plants, and especially that of cotton in African States and in Madagascar as in a number of other countries where the Institute performs its tasks, is by numerous small producers who raise cotton as a supplement to subsistence agriculture. One might estimate more than two million farmers are engaged in the cultivation of cotton in Francophone Africa and Madagascar.[18]

Nevertheless, the I.R.C.T.E. can claim substantial schievement. It has been instrumental in establishing and developing the cultivation of sisal in Madagascar, which harvested about 25,000 tons of fiber in 1970. Between 1946 and 1970 for those African States and Madagascar that are its sphere of responsibility for cotton, acreage under cultivation has almost doubled, but the output of cotton seed has increased from 61,059 tons to 400,000 tons, and output of cotton fibers has increased from 17,000 tons to 148,000 tons. As an indication of quality improvement, the length of the fiber has increased from 7/8'-15/16' to 1 1/6'-1 3/32'.[19]

What can be said, then, concerning these R and D organizations oriented to the plantation culture of export crops? Certainly, they have been pillars of strength for French AID. They have followed a standard pattern of R and D activity. Through seed selection, hybridization, cloning, they developed and made available superior plant varieties, or varieties adapted to climatic variations and resistant to prevailing plant diseases. They have prospected and mapped cultivatible areas. They have studied the soils and the means for their protection, as well as the nutritional needs of the plants and the fertilizer practices and irrigational systems as the means of satisfying those needs. They have attempted to develop and improve techniques of cultivation, harvesting, storage, and sometimes of production or fabrication of end products and by-products. They have developed and formulated the systems and the specifics for the prevention and control of plant disease and plant predators. They all have ties with academic science, producing and publishing scientific studies. They all have a greater or less commitment to the education and training of specialists in their areas of expertise. They are all nominally autonomous agencies of the French government, operating through contractual relations with clients, mostly in Francophone Africa. Indeed, it is upon the support and approval of these client-states that their prosperity and even their survival depends.

They have all been channels for the in-transfer of information into LDC learning and practice. They have adapted prevailing knowledge to the special needs and conditions of the LDCs. They have themselves contributed to the progress of development-oriented science. Through documentation centers, and in the provision of seeds, fertilizers, fertilizer formulas, vaccines (in the case of animal husbandry), pesticides, and pest-control norms and techniques, they have constituted an essential element of the science infrastructure. And, in some instances, they have participated in the formulation of policy in development planning and in the implementation of development plans.

The effectiveness of these institutions is to be explained largely as the consequence of the following: (1) *Their role is clear and their task is specific.* There is no doubt as to what they are, why they are, where they are, and what is expected of them. They are autonomous in the choice of means, but their objectives are given. They have stable responsibilities to a regular clientele. They continuously learn and are in a position to apply what they have learned. They are not left hanging in an academic ether. And their contribution can be measured and their performance judged to a degree very rare in scientific organizations, and impossible for those agencies that are perpetually engaged, in-and-out, with heterogeneous sets of projects. Political masters at the donor and at the recipient end are motivated to

calculate the value of their contribution and to judge them by their performance. (2) *The need for their contribution is real, and significant achievement
is feasible.* Given the backwardness of African agriculture and the great
needs of the people, a significant and, to the researcher and engineer, a
satisfying, self-fulfilling contribution *can* be made on the basis of existing
knowledge, conceptualization, and analytic technique by competent, con-
scientious workers without the gift of creativity and without the necessity
for scientific breakthrough. (3) *They are an integral part of an integral
system of decision and action.* Indeed, they link the plans, purpose, and
purse of the donor to the plans, purpose, policies, and actions of the recipient. They are a part of an operational whole, wherein the availability of
financing, the output of information, the planning and organization of
production and of the change in process, and the development of market
outlets are systematically interlinked.

This does not mean, however, that these institutions in their present form
suffice. For AID policy and from the point of view of the institutes themselves, there are problems and inadequacies that require changes in emphasis and in organization. Among these are the following:

(1) The African states remain rural peasant societies, where village
agriculture provides the means of subsistence. First priority should be to
increase the capacity of these societies to feed themselves and, subsequently,
to produce the surplus of food required to support a working population
engaged in developing the physical infrastructure and then in producing
manufactured products in the towns and cities. The *emphasis* on export
crops does not confront this central problem and indeed may chiefly benefit
foreign consumers, exporters, and an indigenous elite.

(2) The institutes have focused on problems soluable by scientific research and through the rational development and application of technology.
But, and particularly when the objective is to reach and benefit the peasant
mass, a cultural transformation is required. In any case, cultural change and
institutional dissolution is taking place. To teach, to guide, to help, where
social transformation rather than technological innovation is at issue, requires another sort of understanding and, hence, a different research emphasis.

(3) Unlike the AID agencies of other countries, those of France open
themselves to scientific workers of any country. And, no doubt, it has been

an objective of French AID to recruit and train black scientists for work in Africa. With rare exceptions, this has failed. Cadres of black African scientists have not emerged. An indigenous science capability has not been developed.

(4) French sentiment in favor of AID slackens as dreams of glory fade. Ties between France and her former territories are attenuated and sometimes torn. Politicians strum the chords of African nationalism and xenophobia, and these new nations seek to escape a dependence on France, which, whatever its possible values, is at least symbolically onerous. With one of its masters increasingly indifferent and the other growing suspicious and antagonistic, the position of the institutes becomes difficult and sometimes untenable. In some instances, permanent centers have been closed, and research teams have been obliged to leave. Institute directors see the handwriting on the wall, and individual research workers look to their own security.

(5) For the optimal use of science resources in promoting the development of exportable plantation crops or, indeed in promoting the development of any kind of agriculture, the organization of research, the dissemination of research-produced information, and the development of a science infrastructure should be regional in character, rather than divided between and bound within scientifically irrelevant political confines of many small nation states. It should be based, that is, on problems common to a set of soils, climates, pests, predators, and plant diseases. It should be organized to facilitate communication and to extend the possibilities of application. Such regionalization of R and D for tropical agriculture has not been achieved anywhere except in the instance of the Rockefeller and Ford foundation institutes in Mexico, in the Philippines, and now in Africa. Certainly it has not been attained by the French, partly as a consequence of French policy and also because the jealousy of the African client-states, who would deny to all others access to information produced by research for which they have paid their penny.

As a partial response to these problems and needs, the aforementioned institutes and certain others have been brought together "au sein du Groupement d'Études et de Recherche Pour le Développement de l'A-gronomie Tropicale, (G.E.R.D.A.T.)." In my opinion, this arrangement which imposes a more direct and centralized political control and supervision of the autonomous institutes: (1) reflects the desire of those committed to overseas R and D, in the light of loosening ties with Francophone territories and the increase of political uncertainties overseas, to reinforce their direct ties with the French government and to achieve a more secure place within the general French R and D establishment; and (2) also reflects the desire of those committed to French AID to change the method of determining R and D priorities, with less weight to be given to the market —i.e., to the demands of the richer African states who can share costs and whose interest is in further augmenting the inflow of foreign exchange—and more weight given to policy-determined goals. Thus, greater emphasis is placed on upgrading subsistence agriculture and giving greater help and attention to the poor countries and relatively neglected territories of Africa. Through a united front and increased capacity for integrated action, the institutes are enabled to relate more effectively to development planning by client states and to policy formulation with the French government. There is a central point for the coordinated response to the pressing needs and demands of clients and of French ministries, and to the requests of international agencies. Moreover, it is anticipated that centralized control will help to develop a more effective and efficient science infrastructure, by turning single-crop research stations into multipurpose centers, integrating the dissemination of information and sharing on the overhead costs of the overseas installations—thus by increasing the number of access points, widening the dissemination of information produced through French development-oriented R and D and facilitating its use by those responsible for the multifaceted development process. Through the coordinated use of the available specialists and training and experimental facilities, it will be possible to increase the contributions of the institutes to general training of research scientists, agronomists, and engineers. It will also be possible through this system of decision to achieve a greater regionalization in the organization of R and D activity.

These are indeed the goals, aspirations, and intentions reflected in G.E.R.D.A.T. They are still far from being realized. The expectations and reasons behind the formation of G.E.R.D.A.T. are suggested by the following, which is quoted from the Sixth French Plan:

   The highest priority must be given to adapting France's organization of tropical research in agriculture so as to permit an optimal response to the urgent needs of the time. This requires a centralization of control *(effort de concentration)* and a sharing of resources. . . . This concentration of effort can be obtained by new installations, by making available to one institute service and skills available in others, or by the development of relationships with other specialized public or private scientific organizations in France, with work contracted out or with facilities made directly available.

A documentation center for all the information pertaining to tropical development, so organizing such information as to put it at the easy disposition of researchers in the various fields, would optimize the value of such information and would support economic development everywhere.

While the general education of scientists and technicians as specialists in tropical agronomy may not be its responsibility, the G.E.R.D.A.T. can, nevertheless, very usefully contribute by making the educational capabilities of its institutes systematically available for education and training.

In the African states and Malgache the priority task is to create teams operating out of multipurpose stations to carry out development projects that go beyond the increase in production. The objective of multidisciplinary programs will depend on particular needs and circumstances. [There follows examples of the sorts of programs that should be undertaken in different regions] . . . the G.E.R.D.A.T. must directly involve itself in planning and programming research and its application in French territories.

The G.E.R.D.A.T. in sum should constitute an increasingly more effective instrument of response to needs and requests of AID recipients, able to mobilize its specialized teams, its documentation center, and its training capabilities. It should organize itself to facilitate cooperation with the special agencies of the United Nations.[20]

The Development of an Indigenous Science

The French have been ready, willing, indeed anxious, to recruit and train cadres of black scientists from Francophone Africa to share in the research centers and to man the experimental stations installed throughout Africa. They have failed in this effort. Research teams remain French and West European in origin. How to account for this failure? Beyond its implications for French AID, the question relates to the task of creating an indigenous science.

For the elite black Africans who reach the upper rungs of the European or American educational ladders, evidently more lucrative, easier, and perhaps more important career opportunities are available. That elite can afford to say, "La brosse est pour les blancs." Whoever has been obliged to listen to academics in the United States reminisce about their experience with students from Africa will have heard example after example grumpily retold about Mr. A. K. who was trained as a biochemist and became instead a Chef de Cabinet, or about Mr. U. V. who was trained as a physiologist and became instead the Minister of Transportation, or about Mr. O. M. who was trained as a geneticist and became instead the Undersecretary of Commerce for Foreign Trade. But surely, as these things normally go, better a minister or even a sound and effective politician gained than a competent research scientist lost. There is a plentiful stock of whites to replace the latter, but no whites and few blacks can replace the former. It is, moreover, highly desirable that government officials should understand science and its values, limits, applications, requisites, and the means by which it can be harnessed to the goals of development. What is to be regretted is not that the black Africans have trained as scientists in institutions of higher education in Europe or the United States, but that the training of a scientist teaches him so little about science—not theories and analytic or experimental techniques, specific to a discipline, but about science as an instrument of man, as a social force, as a phenomenon to be developed, organized, used for human ends within its inherent limits.

To judge by the opinions of those who are charged in France with the organization of agronomic research for tropical areas, the question is not simply the existence of other more attractive opportunities for the educated black African, but also of alleged difficulties in teaching and learning. A Directeur Général of one of the aforementioned institutes, who was also a university professor, held that the black African was genetically incapable and mentally unequipped to become a research scientist. This opinion may say more about the professor than about his student; yet it is an opinion that prevails to some degree. Another Directeur Général, a man of impressive record with rare success in developing black African scientists and engineers, took a different position. In his view:

> The question is not of native perceptiveness or even of acquired knowledge. Rather, primarily, it is in the habit of rigorous thought, in disciplining the mind for organized persevering pursuit. Prior to discipline, it is a question of motivation; that motivation and self-belief which is needed to persist in the face of failure and frustration. So far as the Frenchman or the European who reaches this stage of his studies, we can take the discipline and the motivation as given. They wouldn't have survived the prior process of competitive selection without it. Not so with the black African. He has arrived by a different route. To train him to be a scientist, he must first of all be selected for his

character, for his capacity for dedication—and to offset that culture of hierarchy and acceptance, he must be given something worth being dedicated to in an enterprise where he knows he has responsibility and the opportunity for initiative. And he must be trained into the disciplines of learning, a disciplining of thought, which is something else than to inculcate information . . . slower, deeper; few Europeans have the patience for it.

Subsistence Agriculture and the Village Economy

The French plantation owner and the colonial administrator were primarily concerned with export crops. These, after all, were central to the imperialist design. It was to produce crops for export that the Frenchman had left his home and had come such a long way to a strange and tragic land. It was from export crops that such a payoff as there ever might be to the imperial homeland must be found. Export crops were interesting also to the new sovereign governments that succeeded the colonial regime; for from exports came the cash that could buy guns and tanks, machines and equipment, Mercedes and Cadillacs, or whatever the ruling elites might want. What counted most for the vast mass of the peasant peoples was not the crops they exported but the crops they lived upon. Economic development turned on a society's capacity to feed itself decently, and then to provide the surplus of food to sustain those who labored to build the physical infrastructure or in the factories. An R and D emphasis on export crops, helpful though it is, nevertheless misses the heart of the matter. The greater and prior need is to raise the productivity of the village economy, of the peasants and small holders who normally feed themselves and produce the food to sustain society. During the past decade French AID has given greater attention to upgrading the village economy and increasing productivity in food crops through: (1) the Institut de Recherches Agronomique Tropicales et des Cultures Vivriéres (I.R.A.T.), and its twin (2) the Société d'Aide Technique et de Coopération (S.A.T.E.C.), and (3) the Bureau Pour le Développement de la Production Agricole (B.D.P.A.). S.A.T.E.C. and B.D.P.A. are application-oriented, whereas I.R.A.T. is a research agency producing information relevant to and useful in increasing the yields or in extending and intensifying the cultivation of food crops and other outputs such as sugar cane, tobacco, fodder, aromatic plants, and spices, which in Africa are associated with peasant farming and the village economy. So far as its agricultural research is concerned, in its pattern of activities and modus operandi it is like the older institutes, directed to the selection, improvement, and regional adaptation of plant varieties, to the study of soils, of plant nutrition, fertilization, erosion control, protection against plant pests and diseases, the technology of peasant and small holder for cultivation, storage, processing, packaging, by-product production. I.R.A.T. research is carried on in laboratories and centers made available by recipient governments (which may constitute a constraint on long-range research programming). It operates a system for testing its research results over a wide range of ecological conditions.

Since 1968 I.R.A.T. has turned its research focus on the totality of village operations and life because the technological potentialities uncovered through research can be realized only in the context of a social organization. Technological innovation normally requires or produces social change, and therefore may threaten existing modes and structures of life. For the traditional village, its full effects can be disastrous. The traditional village as a system for decision making and the organization of activity is not capable of responding to or assimilating all or even many of the advanced technologies that science offers. In this respect, the task of I.R.A.T. differs fundamentally from the older export-oriented institutes. The latter, at least in large part, offer information and propose solutions (in Weberian terms) to *rational* entities organized to respond to opportunities and to assimilate technologies that offer the potential for profitable change. It is therefore incumbent upon I.R.A.T. or any agency undertaking to devise and promote technological innovation in the traditional African village to conceive its task as an aspect of *social* reorganization where a way of life must be changed to suit technology or technology adapted to a way of life. Therefore, in cooperation with the government of Senegal, I.R.A.T. set up in 1968 two experimental units, each including a number of village cooperatives, intended to supply a: (1) laboratoire agro-socio-économique dans le milieu d'application and (2) noyau de progrès technologique extrapolable à l'ensemble d'un territoire.

These represent an attempt to design and test alternative "bundles" of activity, integrated as a way of living and learning, working and earning, producing and consuming, incorporating a more productive technology, yet acceptable and workable in the rural economy.

I.R.A.T. research finds its way into practical application through several channels. Institute activities are undertaken under the auspices, and are carried out in direct conjunction with—hence carry directly into the development, planning, and programming of—recipient countries. I.R.A.T. personnel are sent on numerous missions as consultants and specialists, undertaking surveys and assisting in the development of policy and programs. I.R.A.T. publishes its research results in journals and monographs designed to reach academics, field workers, and policy makers.[21] And I.R.A.T. works with and through its twin action-agency, S.A.T.E.C., in the design and implementation of projects.

S.A.T.E.C. plans and implements projects for the development of "agriculture, artisanal activities, and light industry,"[22] with emphasis on on-the-spot training of extension workers or producers directly.

S.A.T.E.C. and the B.D.P.A. are highly mobile, without clear and continuing responsibility except to fight fires where they burn, the management of crisis, responding to a moving kaleidoscope of demands. They are ready to organize all sorts of change, top to bottom—analyzing feasibilities, planning operations, negotiating the financing of operations, organizing training, managing production or marketing, installing control systems—at the level of the farm, the factory, or the region. Consider some of the S.A.T.E.C. activities.

For a decade it has been engaged in upgrading sugar cane culture in Martinique, Guadeloupe, and Réunion, to "insure the economic rehabilitation of small- and medium-sized sugar cane planters" through: (1) the design of land reforms that involve 1,400 new planters, (2) advisory work designed to improve productivity, and (3) assistance in the formation of planters' cooperatives. It has engaged in surveys and studies intended to carry the benefits of its experience in sugar to other countries such as India, Brazil, Senegal, and the Central African Republic. All this has been in close cooperation with I.R.A.T. Since 1964 S.A.T.E.C. has been engaged in the design of projects or their implementation for the promotion of rice culture in Madagascar, Cameroun, Senegal, Zambia and Cambodia—in close liaison with I.R.A.T. In Guadeloupe, Martinique, and Réunion, S.A.T.E.C. has established a system for land resettlement, intended to develop a new class of small holders through the use of commonly held modern equipment, availing themselves of advanced methods through marketing cooperatives. S.A.T.E.C. has also participated in regional development programs, following the TVA model. In establishing deep-sea fishing operations, it has prospected for fishing areas, designed boats and equipment, planned port installations, storage facilities and processing plants, assisted in the establishment and technical development of private fishing companies, trained fishermen, undertaken market research, and organized marketing systems. It has engaged in numerous surveys, project designs, and training and servicing programs to establish and promote the technological and organizational advance of small enterprises. It has designed and engaged in numerous land reform and resettlement schemes, in irrigation, hydro-agricultural and projects of many other sorts.[23]

S.A.T.E.C. was organized in 1956, and I.R.A.T. in 1960. Both have the same Directeur Général (M. Francis Bour). Between 1961 and 1969 I.R.A.T. nearly tripled in size. In 1969 it employed 140 senior scientists, 75 technical research assistants, and had an annual budget of 37 million francs. S.A.T.E.C.'s staff included 125 agronomists, 34 engineers, 82 senior management executives trained in finance, economics, and accountancy, 65 technicians, with 2,500 field workers and more than a hundred specialists and executives employed by recipient countries as part of S.A.T.E.C. projects.

The B.D.P.A. parallels S.A.T.E.C. in purpose and organization. In 1968 it had a budget of 46 million francs and a permanent staff of 352 scientists and technicians. In the eight years between 1962 and 1969, its budget had increased from 11 million to 48 million francs, and its effective personnel from 195 to 467.

S.A.T.E.C. has a research-oriented twin. B.D.P.A. does not, and for that reason perhaps it has less the aspect of an integral activity and serves more in the role of an entrepreneur of French talents, drawn from many institutions and walks of life to work on special projects, as currently in regional planning for Argentina.

But among B.D.P.A.'s current assignments, there is one of quite extraordinary significance, namely the planning, organization, and management of the new Institute of Agricultural Technology at Mostaganem where 6,000 Algerians will be trained as agricultural engineers to carry out the massive transformation of Algerian agriculture planned for the 1980s. Whether the project will fail or succeed cannot yet be foretold, but the B.D.P.A. certainly has already broken new pedagogical ground in: (1) the determination of the minimum level of prerequisite knowledge and the testing of innate learning capabilities as a basis for the selection of students, (2) the rethinking of training objectives and the specification of the conceptions and techniques to be mastered, of the information and skills to be acquired in relation to task requirements and role potentials, (3) an unprecedented integration of field work and formal learning, (4) the development of study groups as the building blocks of the educational process, each autonomous in the charting out of its learning path, (5) the innovative use of audio-visual aids. So innovative and important does the B.D.P.A. consider its uses of audio-visual techniques that it has organized with French National Radio and Television (ORTF), with universities in Toulouse and in Paris, and with manufacturers of electronic equipment, a new Institut de Recherche et de Promotion Audiovisuel, for the further development and aggressive promotion of such audio-visual techniques.

General Science and Development

Beyond the organization of R and D in support of agricultural specifics and of agencies that bring to bear a range of scientific capabilities for projects in response to recipient needs, France has also organized a part of its general (fundamental, basic) science in support of economic development, i.e., a set of sciences that are related to no particular output or project but provide information and services of general value to the recipients of French AID.

The organization of general science in support of development was proposed, and a scheme for so doing was formulated in the "Congrès pour la recherche scientifique dans les colonies" in the period from 1931 to 1937. It took institutional form in 1943 as the Office de la Recherche Scientifique et Technique Outre-Mer (O.R.S.T.O.M.).

In 1971 O.R.S.T.O.M. employed a total of some 4,000, including 691 research scientists and research engineers, 1,162 technicians, 218 as "personnel d'administration," and 817 as "personnel d'exécution." It had a budget of 110 million francs (as compared to a total personnel of 1,770 and a budget of 40 million francs in 1963). In 1971, besides its centers in France, it maintained twenty-six permanent establishments overseas, including major centers for multidisciplinary teaching and research, and managed major centers of three African countries. Its sections are grouped into four major divisions: (1) earth sciences, (2) sciences of the sea, (3) biological sciences, and (4) sciences of man. To the work of O.R.S.T.O.M. in the general sciences should be added the medical research and services of the Pasteur Institutes, which will not be discussed here, and of the Institut d'Élevage et de Médecine Vétérinaire des Pays Tropicaux, to be described later.

Besides the values of the information outputs of its research, O.R.S.T.O.M. constitutes an important component of the science infra-structure, providing centers for documentation, soil analysis, pest identifi-cation, and the design of pest control systems and other science-based ser-vices. It is also a highly evolved educational institution with full-scale grad-uate programs that have, between 1945 and 1971, graduated 1,017 research scientists with diplomas in various fields of O.R.S.T.O.M. specialization.

The question arises: how is general, or "fundamental," science to be organized in support of economic development?

It must be understood that what the French call "fundamental" science in this instance is not at all the same as what the National Science Founda-tion in the United States means by the term. What the National Science Foundation means by "pure" or "fundamental" science (and the criterion by which it accordingly excludes from its support all research that is not, by definition, pure and fundamental), is, specifically, research that has no practical value and is without discernible relevance to the interests of so-ciety or the affairs of men, i.e., academic research, with problems that follow the academic mode. O.R.S.T.O.M. science, on the other hand, is neither purified of practical values nor are its problems freely chosen. While its values are general (in contrast to the crop orientation of the agricultural institutes), its problems are selected very much with practical values and potential relevancies in mind, and its many strands are integrated into an encompassing plan that is intended to promote economic development. From its incipiency O.R.S.T.O.M. followed two major avenues of inquiry: (1) inventorying (specifying, measuring, evaluating) the elements and the resources, i.e., the climatic, biologic, natural parameters of decision and development in the vast African continent; (2) searching for the means of eliminating or controlling the diseases that afflict man and beast in Africa, through a greater knowledge about their vectors.

O.R.S.T.O.M. has gone far in its inventorying task. Consider some of its achievements.

Taking Inventory

Soil Resources. In 1945 an inventory of soils in Africa, Madagascar, the Antibes, and Polynesia did not exist. By 1970 O.R.S.T.O.M. had developed a highly complex classification system that included 300 soil-utilization categories. The soils of the whole African continent by then had been mapped at the scale of 1/5,000,000 (only for Australia has there been a parallel continental accomplishment). The soils of Francophone territories had been mapped to a scale of 1/1,000,000 with those of the Ivory Coast, Senegal, Upper Volta, Chad, Cameroun, Madagascar, Antilles, and New Caledonia mapped to a scale of 1/500,000 or 1/200,000.[24]

This inventorying of the soil requires more than an immense systematic gathering of facts. If the qualities and character of the soils are to be specified, this must be done in relation to potential uses. Beyond the soil map is the "carte d'aptitude," which designates the type or types of cultivation appropriate for an area under various assumptions, e.g., with irrigation or without irrigation. Hence, the soil map and the "carte d'aptitude" embodies a matrix of hypotheses concerning the physical character of soils, the climatic environment, the biologic parameters in relation to the economic potentialities of land areas in agriculture, forestry, or grazing.

An initial inventorying and mapping of the soils establishes the reference base for the study of future change in soil structure; and the accumulated observations subsequently made provide the data base for the dynamic analysis of the soil formation and deformation. This knowledge of the soils and of their dynamics is clearly relevant to development planning and to the formulation and application of measures to upgrade or protect this ultimate resource.

Water Resources. As with the soil, a major thrust of O.R.S.T.O.M. research has been to gather in and systematically organize the pertinent information concerning the rivers and inland waters of Francophone Africa. This information can only be acquired through a continuum of observation and measurements carried out for a long period of time, covering the slow but recurrent phases of water flow. Based on the recordings of a thousand hydrometric stations, a first synthesis of hydrological data was prepared by O.R.S.T.O.M. in 1956, with a comparatively complete coverage for Francophone Africa anticipated by 1974. Thus, information has been gathered, systematized, and made available that is relevant to and useful for the planning and programming of irrigation projects, bridge building, dam construction, for hydroelectric projects or flood control, and so forth.

Resources of the Sea. For oceanography, too, the task has been primarily one of gathering and systematically organizing facts of practical relevance concerning offshore or sea areas cojoining or of interest to Francophone territories. Thanks to its work, the ocean currents that bathe the continental shelf off the west coast of Africa, and the flora and fauna of the sea that are to found there, are now relatively well-understood. On the east and west coasts of Africa, fishing grounds have been located and charted, and fish stocks have been measured as a basis for determining safe levels of exploitation.

> Research is oriented to the study of fish and marine life in relation to its milieu with a qualitative and quantitative inventorying of flora and fauna as its object; including the determination of useful species, knowledge of their spatial distribution, and their biology, of the appropriate modalities of their exploitation and the extablishment of limits beyond which such exploitation becomes dangerous.
> . . . so far as the principal fish species exploitable through trawling on the high seas are concerned, O.R.S.T.O.M. has provided the solid factual basis for a coherent and rational regulation of fishing in these areas.[25]

Biological Resources. O.R.S.T.O.M. has also engaged in inventorying the flora and fauna of the African earth, preparing *cartes de la végétation* to the scale of 1/1,000,000 and 1/500,000 principally for East Africa and Madagascar.

Science After The Facts

It is surely clear that research can produce useful information concerning the advantages and constraints of nature and other parameters of choice, and in this way general science can usefully relate to practical affairs. Nor can it be doubted that O.R.S.T.O.M. has done yeoman service in helping Africa catch up on the facts about itself, facts that are normally available and are considered useful in other modern societies.

But once caught up on the facts? Once the inventories are made? (Beyond a certain point, the marginal value of incremental facts and of more finely detailed inventories must decline very rapidly). What remains then for O.R.S.T.O.M.? What then should be the long-range role of a development-oriented general science, and what are the traps and problems that inhere in an effort to develop such a science?

But before considering the logic of a development-oriented *general* science when it goes beyond the inventorying phase, it should be admitted that O.R.S.T.O.M. science is not entirely "general" nor completely "development-oriented."

It is also residual, filling gaps and responding to specific needs outside the scope of the other crop-specialized institutes. Thus, its biologists are, in large part, concerned with fodder crops and those plants that provide food for grazing animals, since these are outside the normal scope of the crop-specialized institutes. O.R.S.T.O.M. biologists work also with those institutes on problems considered general or "basic."

O.R.S.T.O.M. science, moreover, includes those extensions of French research into a tropical locale, even when that spatial extension is entirely unrelated to the objectives of economic development. Thus, O.R.S.T.O.M. geophysicists, albeit their location in tropical areas of the French zone, are in fact strategically placed at stations around the globe in order to study a world system from different observation points. Their concern with, or the relation of their work to, economic development is quite incidental.

These activities aside, a main thrust of O.R.S.T.O.M. research has been to gather and systematically organize facts of a sort considered relevant to economic planning and social choice.

The step after the inventory phase has been to generalize upon and to deduce predictors from regularities, observable in the mass of facts, i.e., to produce derivative generalizations that will facilitate the practical tasks of analysis and decision. Thus, O.R.S.T.O.M. hydrology—based upon long, careful studies of the flow characteristics and the secular and cyclical changes of rivers and inland waters—has deduced patterns of relationship within a determined range of variance for, say, water runoff or sedimentation or flow rates as a function of other variables, all useful in prediction and planning.

And beyond the statistically deducible generalization, O.R.S.T.O.M. inquiry seeks naturally to discern, conceptualize, develop, and test causal systems in the explanation of phenomena and process. The soil is a fact; it is also a function of climate, vegetation, fertilization, microorganisms, the movement of waters, the erosion of rock. When O.R.S.T.O.M. pedologists survey and inventory the soil as a resource, they treat it as a fact; but when they approach the soil in relation to erosion, water runoff and sedimentation, degradation, or study the restoration of fertility under green fertilizer or by setting it aside as fallow land, then the soil becomes a system, not a fact. For soil as a fact, classification follows the conceptions and techniques of the United States Department of Agriculture, based on precisely identifiable and quantifiable characteristics, e.g., on chemical composition, which can be easily related to broad survey and utilization studies. For soil as a system, pedologists necessarily turn to the approach of Dokuchaiev, the prerevolutionary Russian soil scientist, who conceived the soil as an ecological function.

In the effort to link the study of water and the study of soil into a system of displacement and change (e.g., in the study of erosion and sedimentation), geology came to play its role in O.R.S.T.O.M. science.

Traditional French oceanography has been the domain of a few gifted explorers who were, in effect, conquistadors, philosophers, artists, and geographers of the seas. With O.R.S.T.O.M., it took a new start with the systematic studies of young, enthusiastic, American-trained teams. By 1963 it had built its first modern laboratory vessel; and its chief, M. Delais, had laid out an ambitious three-stage plan for: (1) gathering the facts concerning the dynamics of ocean currents, the chemistry of ocean waters, the distribution of animal life and biological matter, (2) the analytic studies of these, and (3) the fitting of all the findings into an integrated system that could explain the dynamics and location of fish populations.

A decade later its activities were greatly expanded. Instead of two vessels, it now had seven, and numerous bases of operation, but its enthusiasm had dampened. Gone were any expectations of producing very significant changes in food supplies for Africa. Nor had any general explanatory system emerged. Inquiry had become focused on the alimentary chains in fish feeding. There is, alas, no clear route to be planned beforehand from the study of the facts to the explanation of phenomena.

Struggle Against Tropical Diseases

> In the tropical world, human life is perpetually threatened by numerous and terrible enemies as the agents of disease, that attack men, that attack beasts, and attack the plants that feed man and beast. Add to this the poverty of the soil . . . poor unbalanced diets, and the door is opened to those great endemic diseases that are the curse of the land.[26]

The struggle against the diseases of the tropics that threaten men and limit economic activity has several dimensions: a study of the pathological states of man, animal, and plant, thus to develop the means of curing, ameliorating and developing immunologies against disease, and to administer and apply what is known in human and veterinary medicine and in agricultural practice. So far as French AID is concerned, the Pasteur Institutes in particular bring to bear the force of medical research and practice against the pain and debilitation of human disease in the African tropics and subtropics. Crop-specialized institutes develop and apply the means and strategies intended to protect from predators, parasites, and disease the crops that are their particular concern, and to develop more immune (disease-resistant) varieties. The L'Institut d'Élevage et de Médecin Vétérinaire des Pays Tropicaux (to be described below) undertakes these tasks in the animal domain.

The research emphasis of O.R.S.T.O.M. (microbiology, parasitology, medical entomology) has been on the complex chains by which disease is transmitted, on the ecology of the insect carriers, and on the design and application of the means of breaking the lines of transmission and controlling or eliminating the vectors.

In the first instance, this also has been a vast task of identification, classification, inventorying of the vectors for every region, for every affliction. So also, in the study of the viruses transmitted by insect carriers. To date more than 250 such viruses have been identified, mostly in tropical and subtropical zones.

The habit, habitat, and life cycle of the insect vector, his parasites, and his predators must also be known in order to develop more effective means of control or elimination, through the use of insecticides or larvacides, or more recently, through sterilization and genetic manipulation, through environmental control, through the introduction of predators that prey upon the vectors.

O.R.S.T.O.M. teams have been engaged in every aspect of this research, and in the design and applications of prophylactic measures; public health and sanitation controls besides—but always as one part of an encompassing complex of activities by other national and international agencies. For the great part of its activities, key parameters of control are as likely to be sociological or psychological as biological.

Institut d' Élevage et de Médecin Vétérinaire des Pays Tropicaux (I.E.M.V.T.)

France established its first two schools of veterinary medicine in 1762 and 1795 respectively, and its first school of tropical veterinary medicine at Maisons-Alfort in 1920. The latter was and remains the core of the Institut d' Élevage et de Médecin Vétérinaire des Pays Tropicaux. It replicates and extends the purposes and activities of French veterinary medicine into the tropical and subtropical regions of the French Zone. Since 1920 it has produced 682 doctors of veterinary medicine, thus seeding centers of research and practice in tropical areas throughout the world. It provides various sorts and stages of training, and since 1960 its educational role has been increasing in scale. In the four decades between 1920 and 1960, 532 students passed through its portals, whereas in just a quarter of that time, from 1961–70, the number trained was 722. It operates permanent research centers and laboratories in the Ivory Coast, Senegal, Madagascar, Cameroun, Chad, Central African Republic, Ethiopia, Niger, and has missions elsewhere. It maintains a publication and documentation center. It is heavily engaged in immunological studies, and in the production and distribution of vaccines. It breeds cattle and other bovines and farm animals that are functionally adapted and disease-resistant for the various regions. Indeed, its research runs the gamut of that which concerns the animal maladies and infestation, stock breeding, and pasturage in tropical and subtropical regions.

How Useful Is O.R.S.T.O.M. Science?

By 1971 O.R.S.T.O.M. had to its credit more than 15,000 publications and nearly a 1,000 thematic maps. No doubt it has produced a great amount of information. How relevant? How useful? Of what impact on practice and policy? With the data available, it is impossible to say. Nevertheless, I presume that the great mass of facts that O.R.S.T.O.M. has produced are relevant and potentially useful, far beyond the present capacity of African society to apply and to build upon them. Indeed, the tie into social-political-industrial decision and practice is likely to be the weak link in every system of general science; and for O.R.S.T.O.M. there is a curious lack of concrete evidence of the sort of practical payoff that might have been expected. In oceanography, for instance, the information produced *may* have been useful for fishermen, *may* have influenced national policies in limiting the allowable catch; but clearly O.R.S.T.O.M. oceanography did not lead to the establishment of new fishing industries.

When O.R.S.T.O.M. moves from the accumulation and organization of facts to the development of explanatory systems or theoretical models hypothesizing complex matrices of cause and effect, it leaves the realm of direct social utility and approaches that of academic speculation, where social benefits must be found in transcending established paradigms and conventional modes of thought, to arrive at insights that open new horizons for exploration and discovery. Has O.R.S.T.O.M. produced any such break-through? If it has, the fact has been well hidden.

Camus and Fournier passed this judgment on O.R.S.T.O.M. science:

> It would not do to dissimulate the relative weakness of some of our activities. In 1960 the O.R.S.T.O.M. teams in geophysics, pedology, hydrology, oceanography and medical entomology had attained a satisfactory degree of efficiency and coherence. Not so in the biological and human science, due in part to the recalcitrance of the subjects to scientific inquiry.[27]

Indeed, it is in the study of the organism and especially of social systems that conventional methods of science have been least fruitful, where the subject of study has been most recalcitrant to the probing of research not only for O.R.S.T.O.M., but everywhere in the world. Yet it is in these domains precisely where understanding and the mastery of phenomena is most critical for economic development as society proceeds beyond the sheer exploitation of its fund of resources and the mere extension of its activities.

Development-Oriented Science and the French Science Establishment

It must be kept in mind that these French science-based agencies are a component of a larger public science establishment. Indeed, the development-oriented science agencies of French AID are replicated by normally far larger and more powerful R and D agencies geared to the interests of and to the promotion of economic development in France itself. Notable since the 1960s, accelerating as the decade came to a close and as political uncertainties in black Africa increased and as French dominance there has diminished, has been the pressure and effort of these AID agencies of French science, to find their way back (through G.E.R.D.A.T. or otherwise) to the security of the metropolitan establishment or variously, e.g., through international cooperation, to escape the vulnerability of local attachments.

Aside from these development-oriented science agencies of French AID, other R and D organizations in France, public, quasi-public, or private, also have an important actual or potential relationship to LDC development, occurring spontaneously or elicited through governmental directive. In what follows, some examples of these will be described.

Bureau de Recherches Géologiques et Minières (B.R.G.M.)

The B.R.G.M. is the center of French geological (cum hydrological) studies oriented toward the exploitation of mineral resources and, more recently, toward the exploration for and development of underground water supplies. Its closest equivalent in the United States is the Bureau of Mines and the U. S. Geological Survey, except that besides the information it produces and the public services it renders, the B.R.G.M. is an action agency, primarily engaged in the search for and exploitation of mineral and water resources, working independently, for other agencies of the French government, for other governments, or in association with private corporations. Its role, its raison d'être, is to promote French economic development by reducing resource costs and by increasing the revenues of the French government and the profits of French companies. It is free to chart its own avenues of action, and since it must live largely from its earnings, it is actively engaged in the competition for contracts and for investment opportunities throughout the world.

The first thought of the LDCs—the dream path of their elites to wealth and power—is to find and to exploit mineral resources. Moreover, since agricultural development often depends on supplies of underground water, on both counts the competencies of the B.R.G.M. have been extensively employed in the LDCs both for private interests and for governments.

The work of the B.R.G.M. is varied and complex.

It is, in the first instance, a center for the dissemination of information. In order to provide a French language back-up for geological, hydrological, and mining research and engineering, it offers comprehensive computerized documentation and translation services. For example, 4,500 periodicals are regularly examined and some 3,000 research references (fiche analytique) are furnished per month. It offers also a publication outlet for French reports and research communications on geology and mining. Thus, a component of the science infrastructure for the Francophone world is provided.[28]

The B.R.G.M. also undertakes geological and hydrological surveys and inventories, and incorporates the data gathered in these surveys into geological and geophysical maps and references. Its survey techniques vary greatly from region to region. In France, for example, the B.R.G.M. receives, under legal imperative, rock and soil samples from all new commercial or public excavations and drilling. In the relatively uncharted territories of Africa, it uses aerial photograph and aerial magnetic survey techniques. The B.R.G.M. also offers highly expert service in the identification of the mineral components of soils and ores and in the analysis of mineral structures. Thus, the B.R.G.M. provides data required for one sort of development planning by governments, and for choice and action by private enterprise.[29]

The B.R.G.M. also prospects for minerals on its own account, under contract, or in collaboration with private interests. When it has, on its own account, proven the value of a mineral deposit, it may form a company to exploit those deposits, marketing shares in that enterprise, but continuing to hold a portion of the equity. Thus, the B.R.G.M. has acquired important holdings in major mining developments. It also participates in developmental planning and programming by public authorities. In the instance of one African country it undertook to establish a national plan for the exploitation of the subsoil. In these several respects, the B.R.G.M. actively participates in, and sometimes, spearheads, the developmental process.

In 1968 B.R.G.M. was associated with fourteen syndicates in France for prospecting and for research on mineral resources (lead, zinc, tungsten, copper, iron, fluorine, zircon, rare earths), with operations normally in the hands of B.R.G.M. and with the B.R.G.M. holding twenty to eighty percent of the equity. In the French Zone, B.R.G.M. was associated with seven such ventures, searching for diamonds, copper, bauxite, nickel, phosphate, in charge of all operations, and with forty-five to eighty percent of the equity. It was engaged in similar ventures in Canada, Spain, Chile, and Brazil.[30] In 1968 it was participating in seven companies that had been organized to exploit certain of B.R.G.M.'s mineral discoveries (with equity of eight to fifty percent in each), namely: (1) MIFERMA in Mauritania, producing in 1970 more than a million tons of iron ore; (2) COMILOG in Gabon, producing in 1969 1.36 million tons of manganese; (3) COMIREN in France, producing in 1969 239 tons of copper-content ore; (4) TAIBA in Senegal, producing in 1969 1.0 million tons of phosphate; (5) BOU SKOUR, producing in 1969 1,252 tons of copper-content ore; (6) C.P.C. to produce potassium in the Congo; and (7) SOMIMA to produce tungsten in France. Particularly interesting is MIFERMA, where B.R.G.M. shares control not only with the government of Mauritania, but also with the British Steel Corporation and major steel producers in France and Italy.[31]

The B.R.G.M. also undertakes extensive scientific researches for the purpose eventually of providing information or of developing engineering and analytic techniques, relevant and useful to its practical objectives. Such researches have included studies of very deep earth stratifications, of the conduction of heat through rock formation, of the interrelationships between biological phenomena, e.g., microorganisms and rock formations and studies in the application of probability theory and operational research techniques to prospecting. These provide a science-based back-up to a range of developmental objectives.

Formerly, the B.R.G.M. had acted as a training center, coordinating the advanced studies of specialists from developing countries, and itself offering an internship training program for engineers from developing countries (thirty in 1963). It had also, under the auspices of French AID, been engaged in helping to establish institutions of education in the fields of its competence in developing countries, e.g., the faculty of geology at the University of Brasilia. By 1968, as its role as an agency of AID diminished, the training function had vanished.

The B.R.G.M. had an annual budget in 1970 of 117 million francs ($20 million U.S.). This was considerably less in real terms than its 1963 budget of 100 million francs. In 1970 its permanent staff included 660 engineers and scientists and 350 technicians and administrative personnel, about the same as the number employed in 1963. However, during this period the distribution of personnel as between France, the French Zone (as a function of AID) and elsewhere outside of France, had changed considerably. In the earlier period, about thirty percent of B.R.G.M. effectives worked in the French Zone and less than four percent elsewhere outside of France. In 1969 only ten percent were working in the French Zone, with another ten percent elsewhere outside of France. Of the 27 million francs supporting B.R.G.M. AID activities in technical cooperation in 1969 only 10 million came from the French government. Increasingly, the B.R.G.M. has become dependent on contracts for work outside of France and outside of the French Zone (accounting for 45 million of the 109 million franc total spent in 1969).

No doubt, B.R.G.M. has its problems. Mining is a risky business, and not all B.R.G.M. ventures have succeeded. Caught between technical reverses, political upsets and insecurities, and falling world prices for mineral outputs, B.R.G.M. has shown operating losses (at least during 1968 and 1969) of well in excess of two million francs a year. It has been caught also in a national budget squeeze, for in France as elsewhere there is a rising skepticism and a more critical attitude concerning the practical values of research. And, the position of the B.R.G.M. is in fact an equivocal one. What is it really supposed to do? By what criteria should it be judged? By what rationale, for example, can one justify the work of this subsidized agency of the French government in prospecting for minerals on behalf of private corporations or government agencies in Australia or Canada? Clearly it is engaged in profit making and is embarrassed by its losses. Is it then out to make profit and to be judged by its profits or losses? If so, should it withhold, hide, and on its own exploit the choicest of its discoveries (as a private corporation would) or is it there to open these discoveries to private investment? To French investment? To whose investment?

The B.R.G.M. is or could be a powerful instrument for AID, not simply for the implementation of AID programs (private agents might do that just as well) but in the initiation and planning of such programs from within the political establishment. No equivalent instrumentality of public decision and action exists in the United States. Unfortunately, the B.R.G.M. seems hardly to have played, or been allowed to play, that potential role as initiator and planner of French AID programs.

International Collaboration

Our country-by-country approach may give the impression that science and scientific research is a *national* phenomenon. It is not, though it is squeezed into the procrustean bed of national organization.

Some of the values of transnational collaboration (the B.R.G.M. claims in fact to have established working relationships with the U. S. Geological Survey and with West German Geological Institute or Bundensanstalt für Bodenforschung) might be suggested by reference to geological experience.

(1) In geological surveys and in developing a rational approach to the exploitation of water or mineral resources, the problems faced by the geologists and those concerned with the study and use of the subsoil are not national but regional. The regional nature and supranational implications of geological studies is illustrated in an incident related to me by Miss Delany of the World Mapping Conference. Miss Delany with her colleagues, while engaged in producing a geological map of Africa, noticed a particular rock formation moving from Ghana deep into Guinea. From her experience as a field geologist in Ghana, she recognized the formation as water-bearing, and a water source in Ghana. But the areas where this rock formation penetrated into Guinea was considered to be waterless. The implication was clear: a water source in a desert area. That these observations were unable to spark an exploratory inquiry into the potentialities of water supply for the dry lands of Guinea is indicative of problems of communication and of the perennial lack of opportunity for action or policy initiative by the working scientist.

(2) There is a considerable overlapping and duplication in information and documentation which might be minimized, and a more complete coverage obtained through international cooperation. There are evidently parallel and duplicative efforts in the *British Overseas Geology and Mineral Resources,* in the *French Chronique de Mines et de la Recherche*

Minière and in similar publication by Italians, Germans, and of the French *Bulletin Signaletique* on mining and geology, and the *Geologic Bibliography* published in the United States.

(3) There are as yet unrealized values through coordinated research planning and cooperative studies by national geological research institutes. The study of laterites, a soil formation found in tropical areas, which creates very hard surfaces and is rich in iron and sometimes in other minerals, exemplifies geological problems of great importance to economic development in tropical areas which could be fruitfully dealt with through the coordinated research by the national geological institutes in many countries. Laterites are also of great interest from different points of reference, e.g., to soil scientists and to those whose research locus is on construction and road building.

Bureau Central d'Études pour l'Équipement d'Outre-Mer (B.C.E.O.M.)

The potentialities of science and of modern technology can also be geared into the processes of development through institutions and activities of the private sector. Particularly interesting in this regard is the fraternity of engineering consultants, which has largely grown and flourished under the sponsorship of technical assistance programs. In France a most significant example is the B.C.E.O.M.

The B.C.E.O.M. is a high-powered, expert institution, specializing in: (1) highways and airports, (2) harbors, ports, river clearance and control, (3) railways, (4) town planning and development, including all forms of public utilities, (5) hydraulics and water control, e.g., dams, irrigation and drainage systems, (6) crop storage and processing, and (7) systems-engineering for functional objectives. It is competent enough to study the economic values and the technical feasibility of the projects it undertakes, to organize construction and operations, and to train cadres for future management or maintenance. In 1971 the B.C.E.O.M. employed some 500 French nationals, including about 190 engineers. That year it did a consulting business of 49.0 million francs, producing 267 studies in 67 countries. The volume of its operations has more than doubled since 1960.

In a sense, the B.C.E.O.M. is a private enterprise, aggressively seeking business, like any other. But it also has special characteristics, particularly in: (1) its close links to French planning and policy and (2) its scientific orientation and its efforts to feed back its experience into the body of scientific knowledge and world technology.

The company was established as a Société d'État in 1949 with a small investment (the equivalent of $40,000) by the French government. The informal but close links of French science with French developmental planning are illustrated by the personal history of M. Paul Bourrieres, the General Manager. Bourrieres, a polytechnician, entered the French Civil Service and eventually became Chief Engineer in what was then the Ministry of Overseas Territories, in charge of seaports, waterways, electricity, water resource development, railways improvement, town planning, housing construction in those territories. He later became Chief Engineer of the Overseas Works Services in France. When France divested itself of its colonial responsibilities and powers, Bourrieres founded the B.C.E.O.M. He is also a professor at the Institute of Social and Economic Development at the University of Paris, and is a frequent participant in the conferences and deliberations of international agencies.

Through Bourrieres and his associates (the President of the Conseil d'Administration of B.C.E.O.M. is M. Robert Bonnal, a high civil servant, who since 1961 has been Chef du Service de la Coopération Technique au Ministère de l'Équipement), B.C.E.O.M. has taken on the character of a quasi-public agency. This is not hard to understand. The organization was established on the basis of an investment made and contracts offered by the French government. Through the years, its work has been for governments, and it is absolutely dependent upon their good will. It has indeed been specifically an agency of public policy, far more, for example, than has been the case for such a government agency as the B.R.G.M. Nor is its policy orientation merely a matter of good business. It is ideological as well, assuming an organizational identity that accepts the profit constraints of a commercial enterprise along with the moral responsibilities of the public service and the academic commitment systematically to feed back the lessons and information-outputs of experience as knowledge that should accumulate in the public as well as in the private domain. Thus, the B.C.E.O.M. has arranged for the publication of numerous books based on the field experience of its personnel.

In 1970 the B.C.E.O.M. also established an information center offering, as a free service, to respond to inquiries emanating from French AID agents and agencies or from these engaged in public works in France.

> The extension of its remunerative activities is not the sole preoccupation of this Society. We are as well aware of the moral responsibilities conferred upon us by our origin and by our statutory role. Thus, since June 1970 the B.C.E.O.M. has put the resources of its documentation service fully and freely at the disposition of all the agencies and operatives of French technical assistance. Our documentation service responds at once to any of their queries and makes available to them anything from our repertoire of studies, reports, and publications that they might deem useful.[33]

B.C.E.O.M. is one of a new and growing genre of entities (foundations, public corporations, consultants) that combine elements of the government agency, of the business enterprise, and of the ideologically committed institution, but rarely has one of them formulated so clearly the rationale of its existence or organized its functional activity in the full consciousness of its multiple responsibilities as has the B.C.E.O.M. Thus, the B.C.E.O.M. *idea* might serve as a beacon for others.

During 1968–69, the decade of rapid B.C.E.O.M. advance came to an abrupt halt, attributable to: (1) a diminution of French AID expenditures for the purpose of developing the infrastructure, e.g., building roads, ports, harbors, airports, airfields, flood controls, and other public works, (2) the opening of Francophone Africa to international competition, and (3) a higher proportion of investment by the LDCs in the direct promotion of their agriculture or industry at the expense of public works. Accordingly, B.C.E.O.M. has shifted its emphasis to organizational problems, e.g., to educational planning and programming, to the development of systems for administration and control in the provision of public services and the maintenance of infrastructured facilities, to the promotion of tourism, and to technical and policy problems implicit in eliminating pollution and protecting the environment.

The pattern of B.C.E.O.M.'s support had indeed changed during the sixties. In 1959 all its funding came from the French government. In 1971, while 90 percent of its work remained in the LDCs, only 35 percent of its funding came from the French government, 40 percent was from international AID, and 25 percent from local contracts. At the international level,

German Technical Cooperation and United Nations' projects were particularly important as a source of B.C.E.O.M. funding.

Centre National D'Études et Recherches Céramiques

Though, as now has been shown, there is an elaborate R and D establishment to support agricultural, infrastructural, and mining developments in the Francophone territories, virtually no R and D backup exists for the development of industry in those territories. Yet industrialization is the ultimate objective! Where will be found the accumulation of information and know-how to help the LDCs across that critical threshold?

One possible source is in the cooperative research associations that are attached to French industry, to provide science-based services and to work on problems of general interest to each particular industry. These research associations are supported by contributions from the firms in the industry —but, since contributions are required by law, they amount to support through taxation. The governing boards of the research associations are composed of members of the industry and of representatives of one or more of the French ministries.

In some instances, in response to ad hoc requests, the French government has turned to these research associations as unique repositories of technical know-how in respect to a particular industry for support in its AID programs. Thus, for example, in choosing to honor requests for assistance having to do with ceramic products, the French Government has engaged the Société Française de Céramique to undertake AID missions. In the early 1960s, that Société was more or less continuously engaged in such missions. Often these were in the nature of trouble-shooting consultations, since the factories in developing countries were often established without any locally available competence to provide the research-based information required for their continuous or adaptative operations. The Société had also introduced the use of a small highly mobile electronic device that it had itself developed (adapted from a technique used in the petroleum industry) in prospecting for clays. Because this device is small and light, it is peculiarly suited for developing countries where good roads are few and, hence, where the problems and cost of moving the heavy drilling equipment used in conventional prospecting may be prohibitive.

The involvement with developing countries, unfortunately, depended entirely on the commitment of the then Directeur Général, M. Pol Paquin.

By 1971 he had retired. The public enthusiasm for AID had diminished. The self-interest of the members from industry, who feared creating competition against themselves and who resented any diversion of the energy and time of their research staff, was in command. And the use of the Société as an instrumentality in support of economic development overseas was at an end.

Science pour l'AID

Beyond detailing their organization in France, it has been the purpose of this chapter to illustrate concretely the ways in which the resources of science can be used to support economic development, and to discuss some of the potentials, pitfalls, and problems thereof. Hence the organization of science resources in the AID programs of Great Britain, Germany, and the Netherlands can be dealt with more expeditiously in the chapters that follow.

The systematic use of science in support of economic development has been carried further and has been more rationally organized in France than anywhere else, though even in France it does not reach beyond the protection of health, the development of elements of the infrastructure, and the specifics of agriculture. It should be noticed that, in a time of diminishing AID expenditures, the magnitude of science AID continuously increases—presumably because compared to other forms of AID, it works better, it has proven its worth.

NOTES AND REFERENCES

1. "Gosse Report" (report of a public commission set up to re-examine French AID policies, unpublished in 1972).

2. "Jeanneny Report," Ministere d'État, Chargé de la Reforme Administrative, *La Politique de Coopération avec les Pays en Voie au Développement,* Documentation Française, 1963.

3. Ibid., p. 57.

4. Ibid., pp. 57–58.

5. "Gosse Report."

6. Ibid.

7. "Jeanneny Report," pp. 36–40.

8. Charles DeGaulle, *Memoirs of Hope,* (New York: Simon & Schuster, 1971).

9. Letter (my translation) from Professor Jean Rodier, Chief of the Office of Service Hydrologique, Office de la Recherche Scientifique et Technique Outre-Mer.

10. From OECD, *Development of Higher Education* (Paris: 1970).

11. L'I.R.H.O., *Au Service Du Développement,* (Paris, December 1970).

12. Ibid. (my translation).

13. Letter from Sterling Workman to R. Carriére de Belgarriac, dated 9/9/71.

14. *Sodepalm, Palmivoire, Palmindustrie,* (Abidjan, Côte d'Ivoire, 1970).

15. I.F.C.C., *Rapport d'Activité,* 1970 (Paris 1971), and other in-house publications in French and English.

16. See, for example, the trimesteral revue, *Café, Cacao, Thé,* Bulletins, Hors-series, Ouvrages published by the I.F.C.C. and the A.S.I.C.

17. I.R.C.T.E., *Purpose, Organization, Research Stations, Missions of Technical Cooperation,* (Paris, 1971).

18. Communication from G. Geoffroy Saint Hillaire, Directeur Général of the I.R.C.T.E.

19. Ibid.

20. Secretariat D'État Aux Affaires Étrangeres, *VIeme PLAN, Propositions Concernant Les Credits a Programmer Pour La Recherche Agronomique Tropicale Appliquée* (Juillet 1970), pp. 2–4 (freely translated).

21. I.R.A.T. publishes a monthly review, *Agronomie Tropicale,* and a quarterly review, *Les Cahiers de Agriculture Practique des Pays Chauds,* intended for field officers. By 1970 it had in addition published twenty-two monographs *(Bulletins Agronomiques),* eight monographs *(Bulletins Scientific),* and twenty-five monographs *(Hors Serie).*

22. S.A.T.E.C., *The SATEC Approach to Economic Development* (Paris, 1969).

23. S.A.T.E.C., *Selected References Taken from the General List* (Paris, various dates).

24. Guy Camus, Directeur Général de l'O.R.S.T.O.M., and Frederic Fournier, Inspecteur Général de Recherche à l'O.R.S.T.O.M., "La Recherche de Base au Service du Développement" in *Marches Tropicaux et Méditerranéen*. (Paris, 1971), O.R.S.T.O.M. mimeograph, p. 2.

25. Ibid., pp. 7–8 (my translation).

26. Ibid., p. 2 (my translation).

27. Ibid., pp. 11–12 (my translation).

28. B.R.G.M., *Organisme de Documentation et de Inventaire, des Ressources Minérals et Hydrauliques* (Paris, Janvier 1968).

29. Ibid.

30. B.R.G.M., *Et l'Approvisionnement de l'Économie Française en Matières Minérals: Prospections et Recherches Avec l'Industrie Privée en France et à l'Étranger.*

31. B.R.G.M., *Ses Participation A L'Exploitation Des Gisements Qu'll a Découverts Et Étudiés* (Dec. 1966), also *Rapport Annual, 1965*.

32. B.R.G.M. *Éstablissement Public A Caractère Industriel et Commercial: Statut-Activités-Moyens* (Paris, 1970).

33. B.C.E.O.M., *Rapport au conseil d'administration de BCEOM sur l'activité de la société pendant l'année* (Paris, 1970), p. 2 (my translation).

Chapter 6

Science-Based AID in Great Britain

The Pattern of AID

OF ALL THE COUNTRIES OF WEST EUROPE, BRITISH AID IS CLOSEST in pattern and trend to that of the United States. In both countries, public giving has declined during the 1960s—in real terms quite drastically. Thus, in 1956 dollars, British AID declined from $460.2 million in 1964 to $319.3 million in 1970.[1]

As a proportion of their Gross National Product, British and American AID (0.37 and 0.31 respectively) are close in magnitude—about average for the DAC countries, and far below that of France and the Netherlands. In both the United Kingdom and the United States, financial AID as a proportion of total public AID is relatively large, and the proportion of technical AID is correspondingly small.

The great part of British AID (88.6 percent in 1967–68) is bilateral, and of that 90 percent or so goes to colonial territories and Commonwealth countries, mostly in the form of loans. In fact, British AID has been essentially a continuation and extension of financial assistance to territories administered by the Colonial Office that, after those territories are granted independence as members of the Commonwealth, takes the form of AID. Only after 1958 were India and Ceylon, who had always been outside the Colonial Office domain, given any substantial British AID.[2] That British AID policy is a Colonial Office carryover explains some of its strengths and weaknesses.

"Technical assistance" in 1967 amounted to £33 million, or 18 percent of British bilateral AID.[3] That year, £14.8 million, about half the technical assistance total, was used to pay in part the salaries of British personnel working in the government or public institutions of developing countries. This covered the employment of about 12,000 British personnel. Of these, more than 40 percent were teachers.[4]

What remains is used to satisfy ad hoc requests for: (1) experts on consulting missions, (2) specialized educational or research equipment, and (3) "feasibility reports" prepared by private British consulting firms and paid for by the British government. The British government also provides fellowships for overseas students, and supports educational activities abroad. Finally, a very small part of technical assistance is used to pay for "aerial and land use surveys . . . [and] research costs", i.e. science-based AID.

Before turning to science-based AID, which is our principal concern, a word about bringing students from developing countries to be trained or educated in Britain.

Education in Support of AID

Because English is the great world language, because of England's political, cultural and commercial relationships, and because of the fame of British universities, Great Britain attracts students from all over the world to its educational institutions. In the past, this influx has probably been of direct benefit to the British scientific and academic community and perhaps to the British economy as well, but in recent years, with an expansion in her own domestic education and with the consequent shortage of university places, the rapidly increasing influx of students from overseas is, no doubt, a burden. In any case, in 1964 more than 60,000 foreign students were studying full-time in British universities, colleges, scientific and technical institutions, or as trainees in industry. In 1966–67 that number had increased to 73,400, with 56,000 from developing countries (four-fifths from the Commonwealth). Each year, from 9,000 to 9,500 students are supported by British public funds at a cost (in 1968–69) of £3.8 million.

Their education in Great Britain illustrates the issues raised in an earlier chapter concerning the education in donor countries of students from the LDCs. Clearly, no measures have been taken in Great Britain to redesign curricula so that students from developing countries can be taught what is relevant to their future role and possible functional contributions at home. Indeed, the public or university authorities seem hardly conscious of any need for this, any more than they are in the United States. A part of the difficulty in creating in England a curriculum suited to the role and task of the scientist and technician in developing economies is the lack of any systematic and institutionalized orientation of British research toward the development process or toward the special problems of developing areas— out of which could evolve the subject matter of such a curriculum. The ancient British universities, and the new ones as well, are dominated by the elitist traditions of "pure" science and "fundamental" research, denigrating that which is applied and practiced, so that the generation of students who have come out of those academic centers probably must suffer a period of soul-searching and self-laceration when called upon to turn their efforts to the crass problems of industry and the economy.

In London, Liverpool, and Birmingham, many would nevertheless be drawn, because of the monetary reward and because opportunities are available, into the realm of the applied and the practical, serving thereby the objectives of economic growth. For those who return to Bombay or Delhi, to Cairo or Khartoum, or to Nairobi, there are no such promises of reward to lure them, nor are such opportunities open in industry, agriculture, or government. No vital, progressive, exuberant industry or public activity challenges their ingenuity or offers them an outlet for their talent. If the scientist then is to make a concrete contribution to development, it will be by his own initiative, in the face of resistance. Unfortunately, his academic training, with its emphasis on scientific purity and on fundamental studies, militates against such initiative. Rather it beckons him to bend his efforts not to the goals of development, but to those of Academe, thus to win the approbation of his teachers and of his peers in the European or American university community.[5]

There are exceptions. Training programs for overseas specialists and university institutes which focus on problems of special concern to developing countries do exist. An interesting example of organization of higher education for those trainees is the twin (and overlapping) institutes for economic development and for science policy at the University of Sussex.

The picture has also another side, namely that concerned with the education of British experts to perform overseas the many-faceted technical and scientific tasks of economic development during the coming decades and generations. This requires not only that a viable subject matter be developed and taught; it requires also that there be created the professional opportunities that will attract sufficient numbers of talented men to seek a career in performing the tasks of development. For the latter, it is necessary that the government: (1) decide how many should be trained, for what fields, and for what sort of responsibility and (2) find the means of offering them a sufficiently stable, remunerative, and satisfying career. This last is a problem that has been agitating British academic and political circles. Their general approach has been to create supernumerary posts at certain research institutions for detachable "pools" of overseas experts (e.g., for entomologists at the Commonwealth Institute of Entomology, for plant pathologists at the Commonwealth Mycological Institute, for soil scientists at the Rothamstead Experimental Station).

On Understanding British Policy for Science-Based AID

In the first instance, the scope and organization of science-based AID in Great Britain must be understood in relation to colonial policy and the long-enduring colonial administration under the British; for AID itself is a latter-day extension of that policy and organization.

Science resources were mobilized and were organized for use in response to the demands of colonial administrators throughout a very long, unbroken history of administration under civil servants who were, all things considered, enlightened and sophisticated in matters of science. That experience produced a framework of conception and habit, and an apparatus of control which, whatever its former merit, is now obsolete in mobilizing modern science as a motor force for economic development. The present organization of development-oriented science in France has existed for about two decades; in Germany for less than that. However, in the United Kingdom the organization of development-oriented science took its present form half a century or more ago. Under other names (in 1973), the Overseas Geological Surveys[6] is eighty-five years old; the Tropical Products Institute, eighty years; the Commonwealth Agricultural Bureau, sixty years; the Commonwealth Forestry Institute, fifty; the Anti-Locust Control Center, forty-four.

Consider the basic *conceptions* (of what science is, of what it is for, and how it should gear into developmental choice) that constitutes the framework of *each* of the above-named and the other development-oriented science organizations in Great Britain.

There was the colonial civil service, a tightly welded elite, highly educated in the classical tradition, excellent in administration, who, within their very limited frame of reference, knew fairly well where they wanted to go and were capable of mobilizing the resources at their disposal in order to get there. This governing elite sometimes needed specialized information as a basis for policy and action. To get such information, they turned to experts, and where experts were not easily available, they set up organizations where the appropriate expertise could be developed and the appropriate information gathered. Because new mineral ores were being discovered, they created an organization to analyze and to report on such ores. Because there were indigenous plants that might have local or export possibilities, they created an organization to analyze and to report on the commercial potentialities of locally available animal and vegetable materials. Because an insect pest attacking plantations and farms had to be identified before one could prescribe the appropriate method for dealing with it, they created an organization that stood ready to identify any insect pest.

These science-based agencies, evolved to meet the needs of colonial administrators, fail as instruments for economic development for the following reasons:

1. With a science-based answering service, there must be a sophisticated governing group on the receiving end, who knows what questions to ask of scientists, who knows how to interpret these answers, and how to deduce from the "scientific answers" guidance for practical policy in the context of a special frame of technical and social circumstances. To a considerable degree such a governing group existed in the epoch of colonialism. It exists no longer in the newly independent governments of the Commonwealth, nor in the other developing societies of Asia, Africa, and Latin America. Now those who govern are men of another sort, without the sophistication and skill, and without the education and background that once permitted an easy dialogue between the colonial administrator in the field and the scientific officer in London.

2. The agencies allow no place for science-based initiative in the dynamic evolution of a development program. The sophisticated administrator may ask viable questions of the scientists concerning unusual problems

that arise in the usual frame of social and economic organization. But what of the task of changing fundamentally that frame of social and economic organization? The administrator can, in his own terms, plan certain sorts of change, and possibly may turn to the scientist for information useful to him in implementing his plans. But those who are masters of science or of technology will draw their creative inspiration from a different information base than the administrator's; they will have a different sense of potentialities and possibilities than he does; and they will propose different specifics. When science is valued as a mere question-answering service supplementary to administration, then science can take no initiatives nor make any independent contribution to policy. There is yet another weakness in the British organization of AID that springs from the scientific amateurism of the colonial administrative foreign office official. It is the presupposition that science activity is divisible and self-sufficient on a national, territorial, and administrative-functional basis. Just as has been done in the territories under colonial and post-colonial administrations, each new sovereignty was given its own separate research organizations, as it was given its own fire departments, its police stations, and its post office. Failing to conceive science as a supranational system, no attempt was made to link localized research activity with the science resources and research institutions of Great Britain in order to harness the dynamic force of world science to the specifics of development.

An on-the-spot analysis of the independent research establishments created under the policy of the British Colonial Office in the dependent territories and, ultimately, in new national sovereignties is not, unfortunately, within the scope of this study. But, from the reports of British officials, research administrators and scientists, it would appear that the research establishments in dependent territories, now part of newly independent Commonwealth countries are not doing well at all. On the contrary, many are disintegrating, or have disintegrated, and are on the verge of dissolution. Administrative officials ascribe this to an excessive nationalism, e.g., in the refusal to support regional laboratories, and to a lack of sophistication on the part of government officials. Research administrators and scientists in Great Britain sometimes tell another story. They point out, compared to analogous research in Great Britain, that the research carried out in developing countries is highly academic, lacking development-orientation, lacking practical value, failing to produce concrete payoffs—and that, perhaps, governing officials in developing nations are right in regarding it as a luxury they cannot now afford. Further, professional observers suggest that, at least in many instances, in former days the ostensibly independent territorial research establishments, were actually field activities of British research centers. When independence isolated these research entities, the situation became intolerable for scientists, who, presumably, never had had an outlet into local policy making or practical innovation, and now were deprived of channels to British science as well.[7]

The Organization of Science-Based AID in Great Britain

The waxing and waning of public sentiment in support of British AID is shown not only in the changing magnitude of AID expenditures, but also in the changing structure of its administration. In the early 1960s the Department of Technical Cooperation became the independent Ministry of Overseas Development. And in the early 1970s that Ministry was reabsorbed into the Foreign Office as the Overseas Development Administration.

Running against the trend, the AID expenditure on "research" in Britain has increased from £1.6 in 1962–63 to about £4.0 in 1970–71.[8] This includes allocation to science-based agencies and also grants to liaison offices, university researchers, and so forth. In total, it amounts to less than the annual budget of one of the many development-oriented research institutes in France.

The most significant science-based agencies of British AID are the following:

1. Center for Overseas Pest Research
2. Tropical Products Institute
3. Land Resources Division of the Directorate of Overseas Services.

Others are:

4. Overseas Division of the Institute of Geological Sciences (part of the National Environment Research Council)
5. The Tropical Section of the Roads Research Laboratory (part of the Ministry of Transport)
6. The Overseas Division of the Building Research Station (part of the Ministry of Public Buildings and Roads)

7. The Overseas Liaison Unit of the National Institute of Agricultural Engineering (under the Agricultural Research Council).

Compared to that of the United States (where U.S. AID has developed nothing of the sort), these science-based organizations are impressive enough. But considered of themselves, or relative to science-based AID in France, or in relation to their task, they represent a very small investment and a tiny operation indeed. Four of these agencies will be briefly surveyed.

Tropical Products Institute (T.P.I.)

The Tropical Products Institute (T.P.I.) is the most important development-oriented science organization in Great Britain. Doubled in size during the 1960s, it had in 1971 a total staff of some 360, of whom more than a third were scientists, engineers, and economists. It had been established on the classical pattern to provide a consulting service to the colonial territories on matters relating to the development of export markets for, or to the local utilization of, "renewable" (vegetable and animal) products.

It receives and responds to more than 1,000 inquiries a year coming from all parts of the developing world, sometimes requiring laboratory work or market surveys. Surveys and technical studies are published on matters of more general interest. It claims credit for the discovery of "aflatoxins" in groundnuts, as a carcinogenic agent.

In a sample year the T.P.I. prepared: (1) reports related to the establishment of local food or vegetable product operations, e.g., instant coffee in Ghana, gari in Nigeria, bacon and preserved fruit in Fiji, canned fish in Nyasaland, a food cannery in Trinidad, canned orange juice in Uganda, oil seeds in Antigua and St. Lucia, fiber sacks in Sierra Leone; (2) reports related to export possibilities, e.g., for coffee from Zanzibar, annatto seed from Kenya, sardines, veal, and dried fish from Aden, canned grapefruit and fish products from Jamaica and Kenya, dehydrated fruits and vegetables from Kenya; (3) reports on general market outlook, e.g., for coconuts, drying oils, rami, sisal, and abaca; (4) reports on the economic feasibility of establishing specified operations, e.g., for the general production of particle board from bagasse, from groundnut shells, or from wattlewood and from coffee husks and the feasibility of producing soft-wood pulp in Kenya and in North Borneo. Routine testing is sometimes contracted out, and in some cases the T.P.I. refers problems to commercial consultants.

The T.P.I. operates a center for research on the storage of tropical products at Slough, and a center at Culham, near Oxford, working on a variety of problems, largely related to the design of machines and the development of processes considered of interest to small-scale enterprise in developing countries. One day during the fall of 1971, for example, there was a complex contraption to shell cashew nuts in full clang-bang-chahook-woogle and swing; a wood-fired plant for the distillation of essential oils, a bakery baking nonwheat breads in a study of the feasibility and value of substitute cereal inputs; and the experimental production of press-board out of tropical woods.

T.P.I. experts are continuously on the move on consulting assignments and on visits to areas, activities, and installations related to their research interests and public responsibilities. They average about seventy overseas assignments a year. A stream of overseas visitors coming for consultation or for training find their way through the portals at Grays Inn Road.

The T.P.I., under different names, has been in operation for over three-quarters of a century. It has, what it claims to be, the best library on tropical products in the world. It has, no doubt, developed unusual competencies. As an information and general inquiries center, as an industrial research laboratory, as a body of experts available for consultation, it surely has value. Yet, considering the size of the staff and the great spread and diversity and flux of its efforts, its research must lack profundity and continuity. Its work is necessarily ad hoc, piecemeal. Under these circumstancies it must be difficult for it to follow through on its recommendations or to build upon its experience. It has no power to initiate or to implement projects, or to gear itself into the planning and programming of development. Its leadership knows this and, in the past decade, particularly with the expanded activities at Culham, the T.P.I. has stepped from the inquiries box to move toward the role of initiator and innovator. And more positive links into the processes of world development are in the offing.

The Center for Overseas Pest Research

The Center for Overseas Pest Research was created in 1971 by merging several small units (the Tropical Pesticides Research Unit, Tropical Pesticides Research-Headquarters and Information Unit, Termite Research Unit) with the Anti-Locust Research Center. In 1971 it had a staff of 150, and a budget of £500,000.

The Anti-Locust Research Center began as a small unit to provide colonial administrators with information useful for the control of locusts in Africa. Special circumstances enabled it to escape that passive role and to partake in an international effort to control one of the world's great scourges.[9]

It began with the studies on the locust made by B. P. Uvarov in prerevolutionary Russia. Later, Dr. Uvarov emigrated to the United Kingdom and took up his work at the Imperial Institute of Entomology. In 1928, helpless before the great invasion by the Desert Locust in Kenya, Tanganyika, and the Sudan, the Colonial Office asked that a committee be appointed to advise on locust control in Africa. The research for the committee was undertaken by Dr. Uvarov and a single assistant. In 1929 a small grant was made (£500) to enable him for one year to collect, collate, and disseminate information regarding the desert locust. Uvarov and his assistant brought together information from the British territories and from agencies in French and Belgian areas, systematically charting the location and movements of each of the major locust types—thus providing the primary information required for any rational system of control. In 1931, at the First International Locust Control Conference in Rome, it was recommended that Uvarov's unit, still within the Institute of Entomology, be adopted by all the governments concerned as responsible for centralizing and for coordinating information on the various species of locusts in Africa and Southeast Asia. From then onward, this unit was commonly known abroad as the International Centre for Locust Research.

It was found that the red locust and the African migratory locust were always present in different, quite specific and relatively small areas in Africa. It was from these locations that their invasions always (for unknown reasons) came. On the basis of these findings, it was proposed that control organizations be maintained in the "outbreak areas" to keep a perpetual watch and to prevent any undue increase in the numbers of the locusts. This recommendation was accepted in principle in 1934 at the International Locust Conference in Cairo, but it was not until 1941 that the Red Locust Control Center was set up, and not until 1948 that the Migratory Locust Organization was formed. Subsequently, these two locust species ceased to be a plague threat in Africa. But the greater problem of the desert locust remained, and still remains. World War II stimulated governments to develop locust control units in the field, in order to safeguard their food supplies in strategically important areas. These control units gave Uvarov's unit at the Institute of Entomology an opportunity for action-oriented research and research-based action, preparing the way for much that followed. In 1945 that unit became the independent Anti-Locust Research Center.

Postwar studies indicated that the desert locust could breed and spread from anywhere in an area from India in the east to the Atlantic coast of Africa on the west, from Rwanda and Tanganyika in the south to the Mediterranean coast of Africa, Turkey, Iran, and the Uzbek S.S.R. in the north; some fifty-eight countries are interdependent in the control of the desert locust . . ."the whole invasion area constitutes a great dynamic system in which a flow of locusts took place from country to country and the swarm that invaded one country might have had their origin and breeding in another country thousands of miles away."

Through the F.A.O., an organization for the cooperative international control of the desert locust was created. Subsequently, there were important advances in the use of aircraft and in the effectiveness of pesticides for locust control; nevertheless, "a clear need emerged for the further and wider prosecution of research, not only in locust biology but in operational research on techniques of control and organization. Furthermore, the clarification of the story about the movement of the locusts has revealed that in frequently infested countries it was possible for the locust control resources and equipment to be idle for several months, the idea was mooted of pooling resources so that large control forces could be concentrated in threatened areas. It was also clear that there was a need for training to be undertaken so as to increase the efficiency of national organizations, and for arrangements to be made to facilitate advisory visits by experts to deal with special problems. The nations concerned, under the aegis of the F.A.O., therefore prepared and presented to the United Nations Special Fund a bold and far-reaching scheme called the 'Desert Locust Project,' which envisaged the spending of more than $3,500,000 in six years on surveys, research, training, and operations research."[10]

This Desert Locust Project was started in 1960. It represented the largest single grant to have been made from the United Nations Special Fund. The A.L.R.C., though not the only research unit involved, continued to play an important role. The Desert Locust Project and the French regional "Organisation Commune de la Lutte Antiacredean" manifest "what can properly be called a revolution in the global approach to locust control." A large organization has been created to combat the locust, having at its disposal perhaps thousands of technicians and scientists, fleets of aircraft and helicopters, trucks and ground-spray equipment of all sorts. It is in the nature of the situation that any organization large enough to combat the locust at its peak must be underutilized during periods of lull. Moreover, as control becomes increasingly effective, the apparatus of control will become increasingly underutilized. But cannot it be used for something else also—for the control of other pests as well? In overlapping areas, for example, this control apparatus could be combined with that of CORESTA (Centre de Coopération pour les Recherches Scientific Relatif au Tabac) in its control of the blue mold disease. This apparatus might incorporate other research-training-information centers, to become an international organization for the control of insect pests. It is in the spirit of this potentiality, perhaps, that the Anti-Locust Research Center has become the Center for Overseas Pest Research.

The Land Resources Division

The Land Resources Division has recently developed out of the use of aerial photography to meet the need for natural resource inventories in British client territories. The Division in 1971 had a scientific staff of fifty, including soil scientists, ecologists, hydrologists, agricultural specialists. Beyond land surveys, the Division analyzes the resources and recommends potentials for the development of agriculture, livestock husbandry, and forestry in the land areas surveyed. Members of its staff are seconded to projects of other governments and international agencies. And it has developed a short course in the use of aerial photographs for environmental studies, processing about fifty (1970–71) professionals annually. Its operations are tied into the special competencies available at the Rothamstead Experimental Station, the Commonwealth Forestry Institute of Oxford, and for soil analysis at the Regional Soil Laboratory at Reading.

Tropical Section of the Roads Research Laboratory

The Roads Research Laboratory (R.R.L.), initially a testing station of the Ministry of Transport, was taken over, and developed as a major R and D organization under the Department of Scientific and Industrial Research in the mid 1930s. Subsequently, the R.R.L. found its way back into the Ministry of Transport. After World War II, given the upsurge in road building in dependent territories, the corresponding increase in the volume of inquiries coming to the R.R.L. required that an engineer be assigned responsibility for consultation and liaison with dependent territories, in respect to their road-building problems. It became evident that their problems of planning and building were of another order than those encountered in Britain. Thus, the Tropical Section of the Roads Research Laboratory was created to provide a separate development-oriented unit for research into the problems of, and for the activation of research findings in, those territories.

The Section, though attached to the R.R.L., is financed by the Overseas Development Administration. In 1971 it had a scientific-engineering staff of thirty-five, with Dr. Tingle as Section Head; and was, in addition, empowered to draw nine man-years annually from the range of skills available at the R.R.L.

The staff is organized in four groups that cover the range of the Section's activities: (1) *geology,* with the task of preplanning surveys and terrain evaluation, (2) *engineering,* for road design, road construction, road maintenance, (3) *advisory and training,* organizing short residential courses at the R.R.L., overseas training, and working with the International Center at the University of Surrey, in training engineers from developing countries, (4) *economics,* cost/benefit studies, traffic engineering, safety control.

The Section, beginning with engineering and technical problems, has been drawn by experience and events ever more strongly into economic and social planning. Roads have come to be understood not as technological artifacts, to be evaluated, so to speak, for their intrinsic worth, but rather as skeins of community and the framework for development. As an example, Dr. Tingle told of the conflict in outlook he encountered as a consultant in Addis Ababa on plans to build a road system for that region of Africa. The conventional engineering approach was to determine the shortest feasible (or a low-cost) route between designated end points. But the road system would predetermine the future pattern of land settlement. On it would depend the resources that could or could not be exploited. It would constitute the capacity for integrating activities, mobilizing resources, and bringing outputs into markets. It would set the patterns of regional and international trade. And the road system also would (or would not) bind the scattered people into integral societies and cohesive political entities. To take all this into account in the design of a road system, geared to the social needs and developmental goals of five separate nations, would be infinitely more difficult than to determine the low-cost route between predetermined points. It is nevertheless what is required in our time.

What sort of research is done by the Section? These are some examples: traffic projection, and the development of techniques for traffic analysis and forecasting; collaborating with Ponts et Chaussées in France and other countries to use a mathematical model developed at MIT that brings vehicle operating costs, road building, road maintenance, and other variables into account in computing the "cost of transport," in determining road standards and appropriate highway designs for different regional-cum-economic conditions; the measurement of road deflection with temperature change, to determine overlay requirements; the development of techniques for road stabilization, particularly under circumstances where a crushed stone base is not feasible, e.g., through the use of a precracked cement base; terrain evaluation, and the development of aerial survey techniques and criteria for the analysis of land forms and the characteristics of soil, rock formation, and other elements that must be taken into account in engineering roads and bridges, in providing drainage, in avoiding landslides; and the clarification and systematization of information produced through land surveys.

Beyond its own research contribution, this small group would seem to render yeoman service in searching the biways of technical advance in engineering and construction, for that which serves the purpose and can be adapted to the needs of a wide clientele among the LDCs.

The Future of Science-Based AID in Great Britain

There is no coherent public pressure or interest group to support AID in Britain. Nor does the British public respond to that sentiment of "good fellowship" which has been the Jekyll-side of American great power politics. Nor do they feel the need for *rayonnement* that motivated the French. This, for them, is a time of turning away from imperial pretensions, shucking off the colonial ambitions and responsibilities with which AID is identified. Gone also is the pressure for keeping up with the Americans, for the Americans have retreated even further than they. On the other hand, the ties of mind and culture, of education and experience, of trade and intercourse, of familiarity and friendship developed over two centuries of Empire with the peoples of the Commonwealth and territories, cannot be dismantled by a political act. Living ties will reassert themselves. One way or another, science-based competencies in Britain will continue to complement the striving of those peoples.

British AID has declined drastically. Yet the budgets, staffs, functional activities of the small science-based components of British AID have all increased in response to the need and demand for their services—a witness to their value and effectiveness.

During the past decade, these agencies have increasingly stepped from the passive and subordinate place they occupied in the scheme of colonial administration to act more as initiators of change. And now a further development in this direction is in the offing.

The three largest science-based AID agencies have agreed to a common approach that would permit them to transcend ad hoc consultation and piecemeal servicing, to a start-to-finish establishment of agricultural projects, with new, viable and prospering industries as the concrete end-product of their combined efforts. They propose to act as a team that plans, organizes, supervises, manages projects, or that provides a control nucleus for the planning and organization of such projects. The Land Resources Division would be in charge of the phase of resource survey and assessment, of preinvestment planning and land development, and all preparations prior to the establishment of agricultural operations. The Center for Overseas Pest Research would take charge of establishing agricultural operation, with all the processes of plant selection, protection, and cultivation. The Tropical Products Institute would see to the design and control of the storage, transportation, and processing of crops, by-product production, and the marketing of outputs, as well as attending to prior feasibility studies.

Whatever the eventual fate of this particular scheme, it surely suggests the aspirations of the agencies involved and, therefore, the probable direction of their future development.

The Commonwealth Agricultural Bureaux

Deserving mention at this juncture is a related set of research activities developed at the behest of the Colonial Office, conceived and developed as information services to support practical decision taking by its officials and technicians and the field research they sponsored.

"Commonwealth Agricultural Bureaux include: The Commonwealth Institute of Entomology, the Commonwealth Mycological Institute, the Commonwealth Biological Control, the Commonwealth Bureau of Animal Breeding and Genetics, the Commonwealth Bureau of Animal Health, the Commonwealth Bureau of Animal Nutrition, the Commonwealth Bureau of Dairy Science, the Commonwealth Bureau of Helminthology, the Commonwealth Bureau of Horticulture and Plantation Crops, the Commonwealth Bureau of Pastures and Field Crops, the Commonwealth Bureau of Plant Breeding and Genetics, the Commonwealth Bureau of Soils and the Commonwealth Forestry Bureau. The Commonwealth Agricultural Bureaux are not agencies of the British government and they are not supported through technical assistance funds. Rather, they rely for 80 percent of their income on contributions from member nations of the British Commonwealth. The other 20 percent comes through the sale of published material.

Their primary task is to abstract the world literature in the fields of their cognizance. They also produce card references, bibliographies, and syntheses of current scientific literature. The Commonwealth Bureaux provide an excellent service, producing useful information at a very low cost—but theirs is a service in support of world science and of world research, and not in any particular sense, of economic development. Naturally it is not the LDCs, but the United States, the most technically advanced nation, which is the largest subscriber to their publications.

The Commonwealth Institute of Entomology and the Commonwealth Mycological Institute are also taxonomic centers and provide an identification service for insects and mites in the former, and for plant parasites in the latter. Although this identification service is open to all, the bulk of the inquiries do indeed come from developing countries, suggesting a relative incapacity to make on-the-spot identifications in those countries. This raises an intriguing question as to whether it is best to continue centralized identification services, very distant from the field of action or to build this capacity into the local capabilities. (It is a question as to the best form for a science infrastructure.) To decentralize the identification function would certainly increase the capability in those countries for making direct and exact diagnosis promptly as a basis for quick and appropriate action. A centralized identification serving distant field workers is likely to be justified as economic, when the volume of local inquiries is small, and inasmuch as: (1) identification must rely on unique reference collections of specimens or (2) the task of identification requires an unusual and specialized scientific knowledge and skill. Given the present organization of pest, insect and parasite (and, perhaps, many other sorts of) identification, the need for reference collections of specimens and for unusual personal skill and knowledge, drastically limits the capacity for decentralizing the identification function. However, this "organization of identification" (as other elements of the science infrastructure) has evolved from very small beginnings, but is now expanding rapidly into a quite substantial enterprise. Its evolution has been in the haphazard, random response to ad hoc need, and under those circumstances it is to be expected that these services are open for a considerable rationalization and reform, possibly by applying what has been developed elsewhere to improve technology or to reduce the diseconomies of decentralization. For example, with modern photography and stereo-photographic projection technique, film could be made cheaply available to any research station, field activity, or university in the world to serve as an adequate equivalent to the reference collections of insects and mites at the British Museum or of plant parasites at the British National Collection of Fungal Cultures. This would remove one barrier to direct identification. It might also be feasible systematically to routinize the process of identification so as to eliminate the need for special skill by trained scientists. Taking a page from postwar achievement in chemistry, objects could be classified by reference to structural characteristics observable by the field worker, so that he can specify them on a punch card, with identification and information concerning the identified object, e.g., of the plant or the insect or the fungus, to be retrieved electronically.[11]

AID-Related Science in Britain

We have explained science-based AID as an offshoot of the colonial administration and the imperial tradition. But such AID is also an expression of science policy, i.e., of the way in which the general potentialities of science have been understood, of the way in which research institutions have been organized and developed, of the way in which scientists have been educated, and oriented, of the purpose for which they have been employed and the responsibilities they have been given.

At the simplest level, science policy relates to AID inasmuch as, aside from those small operations within the scope of the Overseas Development Administration, the general fund of science-trained manpower and research capabilities in Great Britain are potentially useful and possibly used as an instrument of British AID.

Before trying to explain British science policy, from whence it has come and into what it is developing, first let us look at the great and long-established research institutions in Great Britain, whose work is related to domestic economic growth in Britain and hence which might in some sense be related also to economic development overseas. To what extent are their competencies and research powers being tapped for the purpose of technical assistance? In this regard, two sorts of scientific institution might be considered: (1) national research laboratories and (2) cooperative industrial research associations, supported in part by the government and in part by the voluntary contributions of private firms.

In order to study the actual and potential relationship of national research organizations in Britain to world development, a number of them were visited, including the Rothamstead Experimental Station in Agriculture, the Hydraulics Research Station, the Long Ashton (Horticultural) Research Station at the University of Bristol, the Fisheries Laboratory at Lowestoft, the National Engineering Laboratory in East Kilbride. These were selected because their research focus seemed of relevance to problems confronting developing countries. From these visits the following emerged:

1. The research outputs of these activities are sometimes directly or indirectly related to the processes of economic development. The Hydraulics Laboratory might test the model of a dam to be built by a British engineering firm in Africa. The National Engineering Laboratories might have contributed to the design of the turbines used in that dam. In this sense, the research activities of national science activity in Britain, just as of world science generally, ties into the process of economic development.

2. Through personal contact and private initiatives, sometimes supported through supernumerary posts to allow government to call on experts for consultancy tasks, or by grants-in-aid from the Overseas Development Administration, these research organizations take an occasional development twist. As a consequence of colonial association and of the long-standing influx of students from British dependencies, British scientists have many personal contacts and private ties and a knowledge of and an

interest in the problems of developing economies. This personal interaction is already greatly diminished and continues to diminish through time.[12]

3. There is no systematic effort to gear the institutional capacities of these national research laboratories and experiment stations into the processes of economic development, or to use them as instruments of technical AID.

4. No opportunity appears to exist for the scientist and research administrator in these agencies to take the initiative in proposing technical assistance policy, or in suggesting the measures which might best be taken in order to put the special competence of particular national research institutions to use in fomenting development.

The cooperative industrial research associations, covering virtually every branch of British industry, were established after World War I at government initiative, and with government financial support. As time passed and presumably, as the value of their service was recognized by industry, the government's share in support of these activities has been reduced. It has been the industry members, not the government representatives, on their Boards of Directors who have exercised de facto control.

Similar research associations exist in other countries. But, measured by the length of their continuous existence and by the resources at their disposal, the British Research Associations are outstanding.

To find out whether and how the British Industrial Research Associations were or might be used as instruments of technical assistance, the following associations were visited: the Printing, Packaging and Allied Trades Research Association, the British Coal Utilization Research Association, the British Paper and Board Industry Research Association, the (now defunct) British Jute Trade Research Association, the Fruit and Vegetable, Canning and Quick Freezing Research Association, and the British Ceramic Research Association. These associations were selected because their competence seemed relevant to programs for the economic development of low-productivity societies.

From this the following appeared:

1. Since radical technological innovations are likely to be disruptive of existing competitive relationships in the industry, research which has such change as its objective will be resisted by some industry membership. Hence, there are intrinsic (though not insuperable) difficulties in the use of

trade association research as a vehicle for creative inquiry leading to radical technological innovation. What the Industry Research Association certainly can and, in the first instance, does do, is to render to its membership a number of science-based services. It becomes part of the science infrastructure, permitting to industry a broader and more precisely understood basis for rational business planning. It carefully gathers the published information, and organizes the statistics useful to business choice. It studies the technical processes of the trade, and measures and quantifies what had before been only intuitively known. It develops control criteria and quality standards. It works on the multitude of daily production problems of breakage and spoilage and spillage and waste. It tests and records the characteristics of material inputs and product outputs. It provides pilot plants to give the members of its industry the opportunity to familiarize themselves with new kinds of analytic equipment or materials or production tools. The net result of all this is that, regardless of its value as a vehicle for invention and for radical science-based industrial innovation, the long-established industrial research association becomes a unique repository of a knowledge systematically organized, quantified, communicable concerning the techniques and practices of an industry. Hence, these research associations constitute an obviously useful instrument of technical education and science-support in establishing industries in developing economies.

2. It would appear whatever contribution these research associations might conceivably make to economic development overseas, that the British government has not recognized the possible value nor tried to make any systematic use of the accumulated knowledge and competence of these organizations as instruments of technical AID.[13]

3. Although scientists and research directors of these industrial research associations seemed sympathetic to the idea that their activities should be given some development-orientation, there are, undoubtedly, company members who would resist and oppose on principle any efforts on the part of their research association to support development assistance even if: (1) the government paid the bill for such efforts and even if (2) no significant threat existed either of the loss of markets or of new export competition being created through such technical assistance. In the face of clear-cut public policy to the contrary, such opposition could hardly be decisive.

Science Policy in Great Britain

On a fine fall day in 1971, in England to complete the research for this chapter, I was visiting at the home of an old friend who is Director of the Science Policy Research Unit at the University of Sussex. Gathered for a talk with us, besides some of his family and his colleagues, were two Labor Members of Parliament from the Glasgow district of Scotland (for the Labor Party conclave was being held at nearby Bournemouth). The MPs were greatly agitated, for just then the government had announced its decision no longer to offset by subsidy the continuing losses and low productivity of the great shipyards of Glasgow, but rather to let them close down. The outcries were loud and piteous—perhaps deservedly so, for the people of Glasgow are poor and unemployment was very high there. And this shutdown promised to deal them a hard blow.

On the outskirts of Glasgow is the National Engineering Laboratory (the British call it Nellie). It was a part of the Department of Trade and Industry and ostensibly its "main function is to help British industry to improve its productivity and profitability." A large, handsome installation, beautifully equipped and well-stocked with scientists and research engineers, it has no equivalent, so far as I know, in the United States.

I asked the two Glasgow MPs, "Have you worked with Nellie, to find the reason for the productivity lag in the shipyards? Has Nellie been used to raise productivity in the shipyards? Why hasn't the government used Nellie to transform the shipbuilding technology in Glasgow?"

All my questions drew a blank. The two labor MPs and the professors of science policy looked at me as though I spoke another language. And the question flicked in my mind: how can I expect this British government to use its established national research institutions and its formidable science resources to plan and promote the technological and economic development of far-off foreign countries when it does not use them, when it has not learned to use them to plan and promote technological and economic development at home?

The same question could be asked about the government of the United States.

The United States and Great Britain have led the postwar world in the magnitude (absolutely and as a proportion of GNP) of their investment in research and development. Yet these two countries have lagged behind the rest in their rate of real per capita growth. A disillusionment with science has become something of the order of the day, and in Great Britain the authorities are in the process of reforming the science establishment.

> The use a nation makes of its skilled manpower is one of the central political questions of the twentieth century, since it profoundly affects the kind of society in which we live. *It is also a major concern for the government which has over the years acquired responsibility for a substantial proportion of the scientists, engineers and technologists who carry out this country's research and development* . . .
>
> *Despite heavy national spending on research and development in Britain we have not profited fully from this investment, for our rate of economic growth has been running at a lower level than that of many of our competitors* [italics mine]. The government has, therefore, over the last few years, concentrated, in developing its policies for research and development, on measures that will most assist industry to improve its performance. We have shifted emphasis from defense research to civil tasks; have reduced the previous concentration on aerospace and nuclear industries—both having very strong defense associations—in order to release effort in support of a wider range of industry; have sought to put more stress on extramural rather than intramural research; and, finally, have given much greater encouragement to the exploitation of innovation.[14]

The situation and the problems of science policy in Great Britain and the United States are superficially quite different, but underlying are the same (or very similar) ideological hangups and barriers to the rational use of science resources. By *ideological,* I simply mean the image that men hold in their mind of how things are and how they ought to be. For we are in a time of ideological transition where the shared and familiar images no longer fit the facts nor can be accommodated to our tasks.

What is this ideological understructure, and what are its institutional consequences?

Government "policy" is conceived as emanating from a body of elected representatives who express the public will. In Britain the emphasis is more on "leadership" and "responsibility" and with us on "representation" and "responsiveness." In Britain, they are longer on eloquence, shorter on muscle. But in the end, in Congress or Parliament, it comes down to a mixed bag of talents and types, with each "member of the Parliament or of the

Congress" operating in a complex of political organizations, dependent upon a clientele of other professional politicians, sharing a single competence in (and with energies more or less totally consumed by) the skills of survival and of manipulation upward through the labyrinthian corridors of perpetual uncertainty in that organizational complex. As the actual tasks of government become no longer routine and residual, but very complex, socially decisive, spearheading change, requiring continuous initiative, creative effort and highly specialized knowledge, those tasks fall further and further beyond the scope and understanding of the elected representative. The deliberations of Congress and of Parliament become less and less relevant to the specifics at issue. "Policy" and its process of formulation becomes more and more distant from the options and opportunities for choice and action. Inevitably, the tasks of choice and control fall to government bureaucrats or "civil servants."

Who are these bureaucrats? In the United States, they are less competent but more open. In Great Britain, they are more competent, but closed— closed upon an ideology that served them well in centuries past but imprisons them now. They are civilized men, educated in the liberal arts, skilled conversationalists, excellent in committees, On the one side, in their conception, are the ministers and party chiefs, who alone bear the burden and responsibility for directing the ship of state, and who make policy. On the other side, are the chiefs of enterprise, who know the mysteries of the market, who possess the key to the "creation of capital" and who alone can properly organize the processes of production. It is for themselves, of the higher civil service, as faithful servants, to carry out policy, following the commands of the ministers, acting as intermediaries and finding formulas acceptable to the masters of the policy and the masters of production.

What happens when the tasks of government are beyond the very limited understanding of the ministers, and outside the scope of parliamentary debate? What when the economic responsibility of the public authority falls outside the scope of the market, and is beyond the understanding or interest of the masters of enterprise? This is the region in which the British administrator now finds himself, blind and groping, regurgitating his old truths, refurbishing time-tried formula, all to no avail. He is without the skills and incapable of the initiative that are now required.

The rate of economic growth and also, since it is the sine qua non of economic growth, the pace of technological advance have become a primary responsibility of every government in the decades since World War II. Politicians crow, businessmen and capitalists congratulate themselves, when things happen to go well. But when they do not? When technology lags and the growth rate sags? "Grow," say the policy makers. "Increase productivity. Grow. That is our policy. We say grow by an annual rate of 4 percent or 6 percent or 8 percent or 10 percent." Such is the policy formulated by the ministers. But how? The ministers have no inkling. Parliamentary debate is quite irrelevant. Nor is the matter resolved by the autonomous motion of the marketplace. The question of "how to grow?" is beyond the understanding of the masters of enterprise. Therefore, the politician and the higher civil servant turned to science.

And what, for them, was science? Their view is encapsulated in Kenneth Boulding's delightful jingle,

> . . . . . research
> Has come to be a kind of Church
> Where rubber aproned acolytes
> perform their scientific rites
> and . . . spend funds they do not hafter
> In hopes of benefits thereafter.

Science, with the capital 'S,' was for them a kind of church, whose high priests might occasionally be brought in to confer with the masters of the government and with masters of enterprise, not to make use of their knowledge (that belonged to another higher, more pure and fundamental realm) but to benefit from their wisdom. Thus the triad: policy makers, entrepreneurs, scientists.

Governments formulate policy. Entrepreneurs organize production. And scientists. . . ? The scientists produce knowledge and invention, hence technology, hence technological advance, which, fed into the market, generates innovation, raises productivity, produces economic growth. All very simple. Since growth was an imperative of policy, and since Science was the source of growth, now Science too had to be accommodated, its gods appeased. And this the ministers and civil servants did by arranging for the establishment of a set of new research institutions, and by opening wide the public purse to university research.

After two decades of trial, Anthony Wedgewood Benn suggests, the ministers and the civil servants have recognized their error. But if creating national laboratories and filling them with decently paid scientists and research engineers and making liberal grants for pure research in the universities does not seem to do the trick, then what will? The ministers do not propose to bone up on science and the high civil servants certainly do not propose to abandon the committee rooms of Whitehall to undertake the nitty-gritty task of actually organizing research and development in relation to (and as part of) a system of technological advance. They do not have the knowledge or competence or inclination to initiate, to plan, organize, control—not that, nor any other complex, dynamic activity. Nor will they bring the scientist in their stead to initiate, plan, organize, and control the activities of science in relation to a system of technological advance, economic innovation, and growth. Not only because those prerogatives belonged to another set of hierarchies, but also because the scientist and engineer are in fact untrained and ill-equipped to step out of the lab into the domain of social decision and public policy to gear their analytic skills into the task of social planning. What is left then?

The British government proposes:

> [The] production organizations [of the Atomic Energy Authority] would be "hived off" into two separate companies—a nuclear fuel company and a radiochemical company . . .
>
> . . . a new Corporation [would be set up] outside the Civil Service to run civil research and development laboratories of the Atomic Energy Authority and of the Ministry of Technology under a single management. The National Research Development Corporation would also be assimilated into this new organization so that its expertise in the exploitation of innovation would be brought directly to bear on the new Corporation's work. The new Corporation would carry out most of its work for both government and industry on a contract basis to insure that its programme reflected the needs of its customers, and that industrial work was carried out in an appropriate environment.[15]

In other words, the British government (in effect, the higher civil service), uncomfortable with and unable to comprehend, organize, or successfully utilize the scientific activities within the public economy, has decided to wrap these activities up in a public Corporation and set them afloat in the market economy. Helped for a while at least by government research contracts, that public Corporation, now outside the responsibilities of government, would be allowed to grow or decline, perhaps to disappear, depending on its ability to sell its service to private industry.

The information produced by the national laboratories operating autonomously having been deemed insufficiently useful, presumably will become sufficiently useful when research is oriented to buyers' demands.

The information offered freely to enterprise was deemed not sufficiently used. It is presumed, therefore, that when information is sold, for private account and monopoly advantage, it will become sufficiently used. Strange presumptions.

The new Corporation is oddly anomalous. What then will *its* objectives and the criteria of its performance be? To advance technology, to accelerate growth, and to serve the public weal? But such services cannot be measured by and will not be recaptured as profits. For the greatest of such contributions there will likely be no payoff in profit to the Corporation at all—and so, by optimizing its contributions to the social weal, the corporation would go under. Or should its objective be to maximize profits and the payoff to those who participate? If so, then the new Corporation will, or reasonably should, patent and hold secret all its juiciest ideas and most commercial inventions, and build a profit base upon these islands of monopoly—like any other private company.

The new Corporation is another gadget, another formula, that would permit ministers and high civil servants to escape responsibility for what they do not understand, and with which they are not equipped to deal. The new Corporation might somehow survive, but it will never be of more than marginal economic or social significance. And the government will have surrendered one of the instruments that it has, though has so far failed to use, to promote the advance of technology and to increase the rate of growth. To use science in support of growth at home or of economic development overseas requires another breed of civil servant with another ideology—conceptualizing the system of technological advance with research as one of its components, and willing to encounter, head on, the task of organizing, managing, changing, and developing that system.

NOTES AND REFERENCES

1. Adjusted by retail price index, *Annual Abstract of Statistics,* Table 306, No. 107, 1970, HMSO.

2. *Aid to Developing Countries,* H.M.S.O., 1963, Comond 2147.

3. *Economic Aid: A Brief Survey,* H.M.S.O., 1968, p. 10.

4. Ibid., pp. 11–13.

5. Dr. H. P. Stout, formerly Director of Research of the (now defunct) British Jute Trade Research Association comments: "I was particularly interested in your comments on the training of scientists from developing countries in Europe and America. There is no doubt that when these persons return to their own country their outlook seems to have been Westernized so far as the scientific approach is concerned and may not be at all suited to the application of science in their own country. It is, as you say, very likely that they will turn to doing academic work which will gain them some merit in the countries where they have been trained, rather than get down to the very difficult application of simple science to their own problems.

Exactly the same thing happens in this country, of course, where the universities unfortunately regard industrial scientific work as being altogether of a lower order, and when scientists go into industry they find it very difficult at first to get used to the idea that whatever work they are doing must have some commercial end in view. It is a great pity that many people do not seem to realize that the problems which arise in the commercial application of science are just as difficult and involve just as great an intellectual effort as the more academic problems. The solution, of course, lies in change in the attitude of the teachers, and this may be very difficult to bring about."

6. Now incorporated into the Institute of Geological Sciences.

7. E. O. Pearson, Director of the Commonwealth Institute of Entomology, enters this disclaimer:

". . . The author's central theme seems to be that technical aid through scientific research, to be most efficient, should be able to identify what problems are of importance to the developing country and within the particular competence of the donor country, that the donor country should organize its own research and training so that these are relevant to overseas problems, and that it should maintain its own organization overseas for research and for implementing the results of research. These are logical if one's aims are the development of the economy of the recipient country and the preservation of a convenient field in which the donor country's technicians can exercise their talents and its businesses market their goods. But they oversimplify the problems. . . .

[He] seeks to explain the characteristics of current British scientific aid in terms of a philosophy of British colonial development that no longer exists, and seems to take no account of the actual way in which British scientific resources were utilized for the development of the dependent territories in the postwar years, or of the political and sociological thinking that has led British statesmen in more recent years (rightly or wrongly) to insist upon making a reality of independence. The official British line that once an ex-Colony is independent, it should be left free to make its own decisions and its own mistakes, and that the U.K. government should not take the initiative in suggesting development policy and resultant aid projects to the newly independent government, nor attach too many strings to the way in which aid funds were used—whether it was due to a genuine and scrupulous desire to respect independence, or merely to a fear of being stigmatized as neo-colonialist —was unquestionably the result of *political* decisions. The present report implies that it was due to the British being too bone-headed to realize that our Imperial world had changed, and to British official and scientific thought clinging to a traditional and outmoded notion of the role of science as an instrument of development. This does less than justice to the views of the majority of British scientists concerned in any way with overseas research, who have consistently disputed recent official British policy because they believed that, whatever the validity of the political arguments in its favour, it would in practice prove ineffective for precisely the reasons set out in this report.

... The French-controlled research establishments in West Africa that are singled out for praise in this report were all antedated, and more than matched in achievement, by research institutes in the former British African territories, founded or fortified with C. D. & W. funds, which have led the way in research on numerous tropical commodities (e.g., cocoa, tea, oil palm, sisal, cotton) and in major medical and veterinary problems (onchocerciasis, arbor viruses, malaria, trypanosomiasis). These research institutes have not been centrally staffed and directed from parent institutes or headquarters organisations in Britain (with the exception of the Empire Cotton Growing Corporation) but they have been vitally associated with the great research institutions of this country because their inception, their policy, their research results and their staffing have been, to a greater or lesser extent, instigated, moulded, scrutinised and assisted, respectively, by representatives of such U.K. institutions operating through advisory committees that have had virtual control of the necessary finances. That such links are now being broken is to a considerable extent due, in my opinion (shared by most British scientists who have been intimately associated, as I have, with the system) to the *political* decision to relinquish direct financial and policy control of such stations. The really vital question which this report not only leaves unanswered, but does not even recognise, is why it is that the French system is acceptable to the African countries concerned, and whether in fact such a system would be politically acceptable in former British territories."

8. *Commonwealth Scientific Committee, Assistance Schemes—United Kingdom,* SC(70)4.

9. Cf. P. T. Haskell, "International Locust Research and Control," The *Times Science Review* (Autumn 1962), pp. 12–14. "The Anti-Locust Research Center," Span. Vol. 5, No. 3, 1962: R. D. MacGuaig, "Recent Developments in Locust Control", *World Review of Pest Control* (Spring 1963).

10. Ibid., pp. 6–7.

11. G. C. Ainsworth, Director of the Commonwealth Mycological Institute, replies:

"At the present stage of development of taxonomic mycology there is much to be said for a largely centralized identification service. The number of taxonomic mycologists is few and the number of undescribed fungi, especially in tropical regions, very many. Centralization does increase the chance that undescribed fungi of wide distribution are described and named once instead of many times as has happened in the past under general decentralization. We here are all most anxious that the objective of decentralization should be kept in view. There are a few hundred species of fungi which are constantly received from all the tropical developing countries and we should like to see arrangements by which these at least were dealt with locally. Handbooks and other aids to this end are under consideration and we should like to help in training the necessary personnel.

"Suggestions to 'routinise the process of identification,' if a bit starry-eyed, are in line with modern fashion but it has always to be remembered that each individual fungus or other microorganism is unique and that microorganisms are as a class characteristically mutable."

As does Mr. E. O. Pearson, Director, Commonwealth Institute of Entomology:

"As regards the report's comments on the identification services of the Institutes, it answers (correctly) its own question as to whether identification services should be centralised. It might be pointed out, however, that the Institute's service is intentionally directed to making decentralisation possible, since the bulk of everything received is returned to the sender after naming, thereby permitting the build-up of authentic local reference collections. The report's statement that much more could be done to reform and rationalise identification and to reduce the barriers to decentralisation by distributing photographically reproduced or microfilmed data springs perhaps from an imperfect knowledge of the subject: a small army of systematists is at work in major museums throughout the world continuously revising the classification of the one million odd species of insects and producing keys for their identification. The barrier to decentralisation is not that we have never heard of microfilms, but that there are so few people at the other end who would be able to use what would be in them. As to the suggestion that the whole process of identification of biological organisms should be reduced to the expression of selected characteristics in numerical terms, thus enabling identifications to be made by feeding sets of measurements into a computer, this is of course the idea at the back of numerical taxonomy, on which a number of systematists are actively working, mainly, however, with a view to the study of infraspecific variation. To suggest that this would eliminate the need for special skills, as the report does, is to fall into the error that besets so many devotees of mechanisation, who fail to realise that vastly more effort would be involved in training field workers to make the very large number of measurements, or numerical codings of qualitative characters, necessary to feed into the machine than in training the small corps of specialists at the C.I.E. It also overlooks the fact that the field worker in, say, the Solomon Islands, having made all his measurements, would still need to feed the data into a computer with a memory capable of accommodating one million numerical entities and decoding each into a Latin name. This would, of course, involve sending the whole set of measurements to a central computer, since the Solomon Islands would be unlikely to have one handy. Might it not perhaps be simpler to continue to put the beetle in a matchbox and airmail it to us?"

12. H. G. H. Kearns, Director of the Long Ashton Research Station (University of Bristol) comments, "I did mention that much of the aid which I had given in the past had arisen from personal contacts overseas in addition to those suggested by the Colonial Office. I also mentioned at the present time, due to the rapid emergence of colonial territories, these contacts had been broken and it appeared that these territories did not know the kind of agricultural scientific aid they required, and it might well be a long time before they were able to select scientific problems that justified outside aid and were likely to give substantial returns in food production. . . . I am sure that the U.K. contacts with tropical agriculture are likely to disappear in a very short time as the cadre of those experienced in the tropics disintegrates. The only hope is to establish some centre . . . that can maintain a vigorous interest in tropical agriculture. . . . Insofar as high level agricultural research is concerned, the obvious link should be between the U.K. agricultural research station and the aided country.

13. C. L. Cutting, Director of Research of the British Food Manufacturing Research Association comments, "As regards Research Associations, I would agree that in general they are perhaps harnessed in many cases to making the best of existing conditions, rather than overturning these with a new set, although there are a number of noteworthy exceptions, which I feel tend to make your generalizations perhaps a little too sweeping. For example in the field of the five food R.A.'s, which are all I can really speak for, the Baking R.A. has in fact developed an entirely novel, mechanical break-making process, which is both quicker and more economical and is therefore being quite rapidly adopted throughout the bread-eating world . . . in my own organization we are also engaged in long-term work on the chemical composition and physical properties of chocolate, which could conceivably, if successful, revolutionise methods of marketing and control.

"Thus whilst I agree with you that the R.A.'s owing to their day-to-day proximity to industrial processes and problems are a "unique repository . . . of techniques and practices," it would be misleading to say that "there are intrinsic (though not insuperable) difficulties in the use of trade association research as a vehicle for imaginative scientific inquiry leading to radical technological innovation" without noting that there are many instances in which these difficulties have been overcome . . .

"Finally, I fully endorse what you say about the British government not having hitherto taken advantage of the R.A.'s "unique repository" in relation to overseas technical aid. . . . We need a more expansive system to provide some built-in freedom for maneuver before we can play our rightful part in the present situation.

"I trust that the criticisms in your report will contribute toward a different attitude towards technology in this country."

14. Statement of Anthony Wedgewood Benn, Minister of Technology, in the publication of the Ministry of Technology, *Industrial Research and Development in Government Laboratories: A New Organization for the Seventies,* HMSO SBN 11 470068 0, 1970, p. 3.

15. Ibid., p. 3.

Chapter 7

Science-Based AID
in The Netherlands

Performance

THE PATTERN OF DUTCH AID IS DIFFERENT FROM THAT OF OTHER countries in our sample. Starting at a very low level and below the DAC average for most of the 1960s, Dutch AID (as a proportion of national income) surged upward from 1968, running against the trend, so that by 1970 it was second in public AID only to France. In the Dutch Second Four-Year Plan, public AID from the Netherlands is scheduled to increase as follows:

Table 9
NETHERLAND PUBLIC AID
(MILLIONS OF DOLLARS)—
($1 = 3.60 DUTCH GUILDERS)

1970 (actual)	196.000
1971 (actual)	269.100
1972 (actual)	291.700
1973 (planned)	319.400
1974 (planned)	368.000
1975 (planned)	423.600

SOURCE: Data for 1971–74 from *The Second Netherlands Long-Term Programme (1972–1975) for Development Cooperation,* (The Hague: Development Cooperation Information Department of Netherlands Ministry of Foreign Affairs, 1971).

Projecting present trends, by 1975 the Dutch contribution in public AID as a proportion of national income will be the highest in the world. This is particularly commendable, since Holland is poor among the donor nations.

Table 10
PER CAPITA INCOME IN DAC COUNTRIES

Country (ranking order in 1966)	National Income Per Capita		Relative Position	
	1966	1970	1966	1970
United States	$3100	$4770	1	1
Sweden	2140	3900	2	2
Canada	1990	3780	3	3
Switzerland	—	3240	—	4
Denmark	1880	3190	4	5
Australia	1660	2700	5	9
Norway	1570	2890	6	7
France	1520	2870	7–10	8
United Kingdom	1520	2170	7–10	12
Germany	1520	3010	7–10	6
Belgium	1520	2590	7–10	10
Netherlands	1380	2400	11	11
Austria	1030	1930	12	13
Italy	940	1700	13	14
Japan	780	1900	14	15
Portugal	380	650	15	16

SOURCE: Data for 1966 from *Development Assistance Efforts and Policies of the Netherlands,* (The Hague: Ministry of Foreign Affairs, 1968). Data for 1970 from OECD, *Development Assistance, 1970 and Recent Trends.*

In the other countries nearly all public AID is through direct donor-recipient bilateral arrangements. Following its four-year plan, the Netherlands intends to avoid such arrangements. Except for its territories in New Guinea, it will channel its AID through: (1) international agencies, (2) consortia set up under international auspices, (3) universities, voluntary associations, and autonomous groups, and (4) the co-financing of private commercial investment.

A great part of Dutch AID is financial, and in 1971 at least $110 million, or 41 percent, was in the form of loans. In 1970 grants represented 63 percent of the Netherlands' "official development assistance commitment," as compared to 73 percent for France, 64 percent for the United States, 54 percent for West Germany, and 49 percent for the United Kingdom.[1]

In order to preserve a specifically Dutch impact, the Netherlands concentrates its AID mainly in Indonesia and India. Pakistan, Turkey, Brazil, Nigeria, Sudan, Tunisia, Kenya, and Yugoslavia have also been recipients.

The Social Context

The organization, the magnitude, and the form of Dutch AID reflects the qualities and the circumstances of Dutch society. With thirteen million inhabitants, compacted into a corner of Europe, the Netherlands is too small, in this age of specialization, to supply the whole range of skills and outputs for the programming of AID. Perforce, it must consider itself as a participant, not as an organizer, in the world organization and the multinational programming of economic development.

The leadership—the educated elite of this old, well-ordered society—are relatively few in number, interconnected, know each other, react, interact, and adapt together in the instinctive way of members of a family. Surely there is something of the family relationship in the whole of Dutch society, cozy and secure, but too restrictive and closely bounded for many Dutchmen. This tradition, habit, and capacity for association—for the quick and effective back-and-forth regrouping of individuals as the pattern of need and priorities change—has shaped internal organization and external orientation of Dutch AID policy.

Though it bears the stamp of historical greatness, the Netherlands now operates only on the periphery of the international power game. It must work through international arrangements and rely on collective security. All this has shaped its AID policies.

Dutch is not a world language, but it is a language, nevertheless, that is spoken by millions outside of the Netherlands, as a consequence of the years of the Dutch empire.

To understand its policies of technical assistance, account must be taken of those years: the century-old relationship to Indonesia and the sudden, complete collapse of Dutch rule there after World War II. Other empires also vanished at that juncture, but in no other case was the loss of comparable importance to the mother country. In Holland, the whole people had become integrated, by private expectations, personal relations, trade, and direct or indirect employment, with those vast, rich, populous Eastern lands. If the loss was enormous, it also came with brutal suddenness, without the warning of long-standing rebellion and struggles for independence that occurred elsewhere. The effects were manifold. Dutchmen, including technical experts, scientists, engineers, specialists and administrators of the highest caliber, were cut off from their work and set adrift. These specialists were generally reabsorbed into the economy of the homeland, where some made a significant contribution to postwar growth. The displacement also left a residual pool of experts qualified by training and experience to work in tropical areas but with no tropical areas left to work in.

The Indonesian struggle left its heritage of bitterness surely, but there remained also strong ties of language, common history, personal contact, private friendships, marriage and family. Since 1964 there has been a political rapprochement between the Dutch and their former possessions.

The Organization of Dutch AID

The Dutch preference for multilateral AID organized through international agencies can be accounted for on several grounds: as a reaction to the experience in Indonesia—to prevent being burned again in bilateral involvements, to avoid the taint of imperialism, and also to simplify and reduce the costs of AID administration. The incremental costs of establishing such bilateral arrangements and the overhead costs of evaluating project alternatives in terms of feasibility and need, of setting priorities, and of administering loans and repayments would constitute a heavy burden upon the operations and would consume a disproportionate share of AID contributions for the Netherlands. This functional dependence gives international agencies a considerable influence upon the policies of the Netherlands and other small countries.

Within the Dutch Foreign Ministry, AID programs are planned by a department for international cooperation which administers financial AID through one branch, and the Dutch relationship to consortia and international agency programs through another. A four-year AID program is formulated and approved by the Parliament. The first of these four-year plans was launched in 1968; the second, in 1972. The first was largely a projection of ad hoc activities. The second is developed around explicit criteria and priorities.

How to account for the increase in Dutch AID at a time of general disillusionment, when AID lags and sags elsewhere? According to my Dutch informant:

> We are Calvinists, heavy with guilt. We see a broken world, and our mission to repair it. Besides, from infancy we have been indoctrinated not with your "Go West, young man" but with "Go abroad, young man . . . go out from this small land." Thus, we reach outward.

Holland, I am told, sends out more missionaries per capita than any other country in the world. In 1971 there were approximately 10,000 Dutch priests abroad, largely engaged in development work. So, too, the Dutch bishops have led in that extraordinary transformation that has released sometimes powerful forces for social change and reform from within the Catholic Church. In the 1970s the one strong pressure group lobbying in support of Dutch AID were the churches, and particularly the Catholic Church pressing for a Dutch commitment that would double the UN proposal that donors contribute 1 percent of their national income as public AID. In 1971 the government started discussion with the churches on the possibility of a cooperative organization of AID programs.

Experts and Projects

In Amsterdam there is a large nineteenth century brick building constructed in the Dutch medieval style, with high, steeply slanted slate roofs, spires and pointed towers, and a flat façade, with tiers of leaded windows. Once the Colonial Institute, now the Royal Tropical Institute, it is indeed an imposing structure, and even more imposing within, where the vast marble halls are filled with treasures from the Indies, temples of rosewood and teak, gods carved in ivory and gold, tributes of kings. Once the gateway to a fabulous empire, now a gateway that leads nowhere, the Institute is an empty, magnificent shell. It has provided a haven for the tropical experts, aging residue of an empire, passing the time between assignments. During the 1950s and early 1960s Dutch technical assistance usually meant experts from this pool answering such demands as might be made by international agencies or by recipient countries. Between missions they worked at the Institute in ways that served to keep them abreast of developments in their scientific or technological specialities, e.g., scrutinizing the literature and in preparing materials for *Tropical Abstracts,* published monthly.

The defects of AID through the ad hoc missions of experts was described earlier. Such an approach permits no follow-up on recommendations, nor any systematic feedback of experience as a basis for cumulative learning and the progressive and disciplined development of science and technique. As a form of technical assistance that depends on the occasional requests of a large and changing universe of recipients, it gives the donor a minimal responsibility and control. It might be said, parenthetically, that "experts," no matter how expert, held over from colonial administrations may be unqualified for the tasks of AID, inasmuch as they are shaped by an outlook proper to a former epoch, but not to this time of economic and social revolution.

During the 1950s and 1960s the Dutch government was under pressure to change this system of technical assistance. Age was depleting the ranks of the experts. The young were not taking their place, reflecting the lack of learning opportunities or of secure and rewarding careers. To offset this depletion, the Associate Expert Program was started in 1954, when four inexperienced young university graduates were assigned as assistants or apprentices to established FAO experts in the field so that they could acquire the experience that would enable them eventually to take their place among the scientists and technicians available for work in the tropics. By 1966 there were 116 graduates in the program. Germany and other donor countries have followed the Dutch example. Further, in order to provide stable careers for specialists, the Netherlands in 1964 instituted a new category of civil servant whose principal work would be in AID missions; when not employed on missions, such specialists would be assigned duties in Holland intended to refresh or to extend their knowledge in their field of science or technology.

In 1964 the form of Dutch technical assistance shifted from the passive pool of experts on call, available for temporary assignment on the requests of recipient LDCs or international agencies shopping for this skill or that, to a system of bilaterally arranged projects, where the donor undertakes to perform a task by a team with a logistic and organizational base in the Netherlands, i.e., to a system of project AID.

The second four-year plan again turned away from bilateral arrangements, but not, I think, to return to the pool of experts. The oncoming organization of Dutch technical assistance is foretold in the conception and anticipations prevailing at the International Agricultural Center at Wageningen (I.A.C.) at the onset of the second plan.

Science-Based AID in Agriculture

For generations the prosperity, indeed the survival, of the Netherlands has depended on the rise in agricultural productivity, hence on the advance of agricultural technology. Higher education in agriculture and agricultural research is centered in the University of Wageningen and in the research and training institutes located there. At Wageningen also was carried on the Dutch research on tropical crops that transformed the economy of Indonesia during the Age of Empire. Subsequently, Wageningen became the operating base for Dutch technical assistance programs in agriculture, with the International Agricultural Center (I.A.C.) recruiting experts for overseas missions, arranging fellowships, and organizing agricultural projects and training programs for students from developing countries. Since 1952 thousands of agricultural extension workers have been trained there with trainees coming not only from LDCs but from Europe as well.

In 1964, with the onset of Dutch "project AID," the I.A.C. began to establish training programs overseas, starting with a cattle-breeding instruction center and demonstration farm in Tunisia. The Dutch emphasis has increasingly been on building up the educational capabilities of the LDCs themselves, rather than bringing students to the Netherlands.

The I.A.C. conceives itself as a catalyst for change and as a nucleus of organizational action. At home, its staff ranges from eighteen to eighty, but it draws upon the whole range of Dutch agricultural skills at Wageningen and elsewhere, assembling teams suited to a task, preparing them for action, integrating the capacities of diverse agencies. In 1971 it had 250 operating personnel abroad. From 1964 to 1971 the magnitude of its operations has grown tenfold.

As envisaged by the I.A.C., the multilateral organization of technical assistance under international agencies will have a new form. No more a mere intermediary in the assignment of experts, nor itself the organizer of projects, it is anticipated that agents of the United Nations or of other international bodies will act on-the-spot to integrate the activities and arrange for the collaboration of national donor teams of experts, gearing these into the work of the LDC's own development planning authority. Such indeed is the emerging pattern of collaboration under international agencies in Indonesia, Kenya, Tanzania.

The I.A.C. further envisages the formation of multidisciplinary regional centers, under international auspices, forming a bridge to European and American expertise, providing a source of research and an instrument for planning, and acting as a positive force for action in those problem areas where the research, planning, and policy functions need to be integrated in relation to regional conditions, e.g., in ecological centers or in urbanization centers.

Education for Development

Dutch experience provides an interesting illustration of the problems of educating students from developing societies.

To bring students from abroad and to educate them in the donor's own institutions of higher learning has advantages for the donor. It enhances its prestige in a deep and enduring way. It creates a web of connections and a community of values in science, politics, industry, culture. Few investments have ever redounded with such benefits to a nation as those which have come to Britain as a consequence of the endowment of Cecil Rhodes to provide scholarships for selected foreign students at the great English universities.

The educational prestige of the United States, Great Britain, and France is so great that more students from abroad seek to come than can be accommodated. Not so with countries like the Netherlands or post-World War II Germany. They have made a special effort to create institutions that would accommodate the needs and attract students from developing societies.

Though Dutch is spoken by the many millions in the former Netherland East Indies, it is not a world language. This has important repercussions on the Netherlands' program of international AID. It has meant, for example, that students from abroad cannot be expected to know Dutch before they come nor to spend much time learning it in Holland. Therefore, foreign students cannot ordinarily be integrated into the regular courses of instruction offered by Dutch universities and schools. If students from developing countries are to be educated in the Netherlands, then a curriculum must be especially arranged for them to be given in one of the world languages, not in Dutch. Therefore, to attract a substantial number of foreign students from elsewhere than Indonesia, the Dutch have had to prepare a special curriculum, offering courses in English or French.

Ultimately, this handicap proved to be an advantage to the Dutch. Because they were obliged to prepare a special curriculum for students from developing countries, they were forced to confront the particular and peculiar needs of that bloc of students. Consequently, they were pressed into evolving a subject matter that would enable trainees to make an effective contribution to the economic development of their native countries rather than, as is done elsewhere, merging these students with a dominant student body from the technically advanced society being prepared for careers in the industrialized economies of the West.

A dozen or so institutes in the Netherlands are specifically designed for the educational needs of developing countries. Autonomous, in 1971 they were subsidized by the Dutch government by about $10,000,000 a year, with another five to six million made available in scholarships.

In its creation of a research-based curriculum designed for the needs of students from developing countries, the International Training Centre for Aerial Survey (I.T.C.) has been a model for the world.

The Institute, proposed through the United Nations, was founded in 1950 under Dr. I. W. Schermerhorn to deal specifically with a science-based technique which, though complex and expensive, is peculiarly suited for geological surveys and mineral prospecting, for the inventorying of forests, soils and pasture land, and for all sorts of mapping in developing countries with difficult-of-access interior regions. It draws its students from technical organizations in government bureaus in those countries. They come to acquire skills that they will use in the work that they are doing or expect to do at home. They are trained in three general fields: Photogrammetry, i.e., the relation of aerial photography to mapping; Natural Resource Inventory, i.e., the interpretive use of aerial photography in geology, soil survey, land classification, soil classification, forestry measurement and inventory; Aerial Photography and Survey Navigation, i.e., aerial photographic production, survey flight and navigation, image analysis and evaluation. Courses, given in English, are from seven to twenty months, and lead to the granting of specified diplomas.

Through its own research, the Institute has developed the technology of aerial survey in relation to the peculiar needs and capacities of developing countries, e.g., manual triangulation for integrating aerial photographs in lieu of computers. Hence, it not only transmits skill and information; through development-oriented research, it develops skills and produces information to be transmitted.

Furthermore, in order to support the application of the techniques it teaches, and specifically in order to back up the work of its former students, the I.T.C. developed a consulting service, sending faculty into the field to evaluate aerial survey programs and policy proposals on the spot, and to solve special problems. In this way, a field experience is fed back to the research-teaching nexus.

Another of these institutes that embodies an innovative idea is the Research Institute for Management Science in Delft (called RVB), which has developed its program and evolved a curriculum designed entirely to teach the management of *small enterprise*. Originally, it was intended to reach a Dutch clientele, but the small entrepreneurs did not respond. They were not interested in being taught by academics and professional engineers; instead, father to son, they trained their own succession in their own shops and factories. In 1955 the Institute found its niche in training those concerned with management in developing countries. It combines formal instruction with supervised field work in selected businesses, capped by an independent, analytic plant survey by the student. By 1971 the Institute had graduated 650, and was currently training 120. It had established operations in its own image in Manila in the Philippines, in Buenos Aires and Rosario in Argentina, in San Paulo and Recife in Brazil.

It would take a more intensive study than has been possible here to evaluate its program or to assess how well the RVB has done its work. It is, however, a unique response to a crying need. Business schools in the United States or elsewhere do not teach the student to start, to organize, to manage his own business. They do not teach him to be an entrepreneur. On the contrary, the business schools have taught their graduates to occupy one of the specialized niches in a large corporate organization. As for the skills of entrepreneurship, these are to be learned through father-to-son apprenticeship or, at least, through observation and indoctrination in a business family environment. What of those who are not of an entrepreneurial clan, nor raised in a business family? What of those classes or minorities outside the informal channels of business learning? How, for example, is black capitalism to be established in the United States? And in any society, when small entrepreneurship and the management of small enterprise is not performing well enough, how is it to be upgraded?

For both, to upgrade the management of small enterprises and to open the path of entrepreneurship for those now outside the gates, some formal system of training is necessary. The RVB opens a possible way. It should be seen as a pilot project in the development of a new dimension of business training.

The International Corporation and the Dissemination of Advanced Technology.

The Netherlands has given unusual attention to developing a system for promoting the extension of Dutch enterprise, into developing countries.

> Cooperation with private business in the form of "starter projects" is aimed at encouraging the establishment of Dutch firms in developing countries. . . . Three types . . . (1) exploratory . . . consisting of the collection and analysis of data required by Dutch firms for investment decisions; (2) limited infrastructure projects comprising investigations into such fields as training, transport, housing, etc., necessary for the proper implementation of investment plans; and (3) pilot projects. Expenditure . . . shared between the government and the private firms. . . . Government has been requested to bear between 59 to 80 percent of total costs, . . .[2]

The Dutch government has also taken steps to: (1) guarantee the investment of small- and medium-sized companies in less-developed countries against political risks, (2) provide for a share of equity investment, and (3) grant tax and other concessions as incentives for overseas investment. Yet it is not these measures that explain the extraordinarily high out-flow (as a proportion of national income, the highest in the world) of private investment from the Netherlands into less-developed countries. That is to be accounted for by the importance of Royal Dutch Shell in the Dutch economy. Thus, for example, "petroleum investment" accounted for 64.5 percent and 70.2 percent of total Dutch direct investment in the LDCs in 1965 and 1966.[3]

The Dutch economy includes four massive corporate enterprises (Royal

Dutch Shell, Unilever, Philips Electric, and AKU, a chemical cartel) and Statemines, a large public corporation. These stand apart from the others, autonomous, powerful, with a particular and perhaps privileged status. In addition, there are numerous smaller but still very substantial companies and cartels, to whom the government's overseas investment incentives seem primarily to apply.

Concerned with the role of corporate enterprise as a carrier of advanced technology into the less-developed world, I interviewed executives in a number of Dutch companies selected for their interest in the LDCs. From these interviews the following emerged.

1. The four giants apart, the great barrier limiting the initiative of any individual Dutch firm in establishing operations in the less-developed world is that its skills and technology are but one component of an industrial-infrastructural complex, and that it cannot operate successfully elsewhere except as part of such a complex. Hence, more fruitful than guarantees and subsidies would be the simultaneous establishment abroad, coordinated through public initiative, of complementary industrial operations, coupled with public investment to provide the infrastructure required for the co-joined or complementary enterprises.

2. A large textile company presents an interesting and curious case, with implications perhaps beyond itself. The firm was established in the early eighteenth century. Before World War II, protected by a tariff, it produced approximately 90 percent of its output in cotton cambric, destined for Indonesia. After the war it lost its place in the sale of cotton cambric to lower-cost local manufacturers and to Japanese competition. By 1968, out of the cotton cambric trade entirely, it had successfully shifted to the production of high-quality cottons and synthetics for the European and American markets. Still, over the years, it had acquired a range of operating skills and know-how relevant to the establishment of operations or to marketing in Indonesia. And so, to capitalize on this fund of know-how, it opened a consulting division that successfully markets to the LDCs, or to those who might invest there, a range of relevant skills and experience.

3. Philips, one of the world's largest producers of radios, television, and electronic equipment, has established a pilot plant in Utrecht to design and develop production systems adapted to the special conditions of the LDCs. In practice, this means the development of production systems optimal for

batch production under circumstances where complex installations would be difficult to maintain or repair. Managers, supervisors, and sometimes workers from the LDCs are brought to Utrecht, where they participate in the design of the factory organization and are trained in its operations. Then the machines and equipment are disassembled, packed, and returned to the LDC (along with the supervisory and working crew), to be reassembled and installed, with operations carried on as they had been in the pilot plant at Utrecht. By 1971, Philips had established 27 factories among the LDCs.

The Industrial Demonstration

How to introduce the village craftsmen and peasant masses or even city workers in Africa, Indonesia, Latin America, India into a greater familiarity with the machine, its adaptability, its potentialities, its usability? Perhaps we could take a page from the old county fair, where American farmers came together from out of the backwoods to feast and be entertained, to show off their own and to admire their neighbors' practical achievements (with blue ribbons for prize horses, cows, pigs, pies, liquors, cabbages, fruit, and corn), to buy and sell, to examine and try out a range of machines and instruments offered for demonstration and sale. So also traveling shows that offer entertainment and information, that teach and stimulate practical change, could be organized and kept on the move through the LDCs, or could be permanently established as places of pilgrimage. As models, of such industrial education through demonstration, consider the following.

1. In Moscow there is an exhibition center set up in the traditional world's fair style, operating continuously since the mid-1930s. Thousands visit it daily, from every remote corner of the USSR, crowding broad, fanciful avenues lined with towering gilded-plaster statues of heroic workers, peasants, and soldiers, flanked by the mock magnificence of exhibition halls. In hall after hall, the visitors see displayed the latest machines, from those that press wine to those that roll cold steel, from sewing machines to computers. Seeking out what they might use, what interests them, they watch, listen, examine, and, under instruction, try out the machines. Could not such centers be installed as mass-learning devices among the LDCs as well?

2. In Rotterdam the Bouwcentrium (Building Center) offers a permanent

exhibition of virtually all kinds of building and construction materials and household appliances and of architectural innovations and building techniques. In 1964 it maintained a documentation center and an information service for builders, architects, engineers, and housewives—answering 40,000 inquiries per year. Information which could be given from published sources or from the accumulated knowledge of Bouwcentrium specialists was made available to anyone requesting it at the cost of the one guilder entrance fee. When an inquiry required special studies, the client was asked to pay for the needed research. When an inquiry pointed to a problem with important general implications, the Center might start a study headed by a team of its own specialists, always with an ad hoc advisory group, drawn in the characteristic Dutch fashion from representative industrial sectors, technical specializations, and science disciplines throughout Holland. Research studies were variously supported by the Dutch government, international agencies, industrial firms, and building contractors. Thus, the Bouwcentrium, the brainchild of one energetic and visionary individual, acted as a permanent exposition on building materials and techniques, as an information center, and as a consumer-oriented research agency. In 1964 it had a permanent staff of 270, including 135 professionals, and drew upon 500 others in various advisory and consultant capacities. It also offered a six-month postgraduate program for twenty-five students from developing countries, teaching them construction in relation to social planning and functional engineering. Each student worked under the guidance of Dutch experts, in the expectation that his work would lead to an actual project adapted to the needs, and constructed for the benefit, of his own society.

The Bouwcentrium is interesting not only for what it is doing but also because it has evolved an approach of great potential value for developing countries, in giving a technical orientation to the outlook of the general public, i.e., in propagating the cognition of mechanism and of process and of bringing the science infrastructure into the support of choice and planning by households and small enterprise.

3. After World War II the TNO (the Netherlands Organization for Applied Scientific Research) set up a pilot plant for the production of plastics and products fabricated from plastic materials. It made that pilot plant available to Dutch firms for experimentation and learning. On the basis of that experience, a substantial industry was created. Surely, this has implications as a technique for economic development.

Industrial Research for AID

The TNO was established in 1930 and subsequently played an important role in the industrial development of the Netherlands. It is, in fact, a cluster of institutes, some focused on national defense, others on food and nutrition, and others on health, but its major orientation is on industries. In 1971 some 3,000 of the 5,000 employed by TNO were involved in industrially oriented R and D.

In 1964 the decision was made to use the TNO in the organization and implementation of project AID. TNO institutes were canvassed. Project suggestions were forwarded to the Dutch government. Subsequently, TNO teams were assigned project tasks, e.g., feasibility studies, the establishment of a capability for food-quality control in Saudi Arabia, research on the use of raphides in developing a repellent to prevent rodent and insect damage to human food, a pilot plant for producing date syrup in Libya, research on the production of bread from composite flour, experimental fish processing at Lake Victoria in Tanzania.

In 1971 a new office was established to control the activities of the numerous institutes for a larger more coordinated TNO role in AID. This is evidently part of a basic reorganization of the TNO, wherein the various institutes will be vectored in for a coordinated attack on several major problem areas, e.g., pollution and the environment, traffic control and road safety, and economic development through Dutch AID.

The Dutch Consulting Firm as a Bearer of Advanced Technology

The Netherlands Engineering Consultants (NEDECO) is an independent engineering organization that is primarily involved in economic development projects. Though not subsidized by the government, it is so closely integrated with government policies and personnel as to be virtually an agency of Dutch AID policy. This, in some ways, is equivalent to the B.C.E.O.M. in France. As was the case in France, it was founded by a high government engineer in overseas territories after the end of empire had terminated his career in the colonial administration. The complete and violent breakaway of Indonesia, and the virtual Dutch withdrawal from direct overseas relationships that followed, spelled a difficult period for NEDECO. Not until after 1957 did it establish itself as a financially viable enterprise.

What distinguishes NEDECO, as compared to the B.C.E.O.M. and the major engineering consultants elsewhere, is the extreme degree to which it acts as an intermediary—organizing and coordinating the range of Dutch skills and talents drawn from institutions, enterprises, and agencies outside itself and recombining these for a variety of purposes. With a permanent staff of some 30 professionals, in 1970 it was engaged in 54 studies and projects, including 21 on transportation infrastructures, 8 in hydraulic engineering, 13 in agricultural engineering including irrigation, land reclamation and flood control, and 3 economic feasibility studies. In these tasks NEDECO drew upon 17 independent organizations, including consulting firms, university institutes, and public agencies.

It undertakes no operational activities, i.e., no design or construction, but only the staff work of analysis, planning, bid-evaluation, supervision. On its advisory board are representatives of the Royal Institute of Engineers, the Ministry of Foreign Affairs, the Ministry of Public Works, the Ministry of Economic Affairs, the Ministry of Agriculture, the Technical University of Delft, the Technical University of Eindhoven, the Agricultural University of Wageningen, the Central Organization of Research and Applied Sciences, the Netherlands Railways, the Zuyder Zee Works, and the Netherlands State Mines. In addition to the competence of various engineering firms, it draws upon the expertise of all of these public agencies. The manner in which it operates, its actual and potential relationship to the transfer of science-based technology, and its tie-in to Dutch technical assistance programs can be illustrated by the following example.

The most exportable Dutch expertise is in hydrology and water control, related to land reclamation, canal building, ports and harbors, water siltation. The Japanese, I am told, wanted to reclaim certain of their lagoons, and they contracted with NEDECO for its services. The Japanese did not want the Dutch to do the work but rather to guide and teach Japanese engineers to do the job. A brain trust of experts was set up in the Netherlands. Each of these experts visited the area to be reclaimed in Japan and studied the situation. One Dutch consultant remained in Japan as a liaison between the working engineers and the board of experts in Holland. Thereafter, all studies, plans, and reports of results were forwarded to the Dutch, were examined and evaluated by them, and returned with criticism, questions, suggestions, advice. When the job was completed, Japan had reclaimed its lagoons and also had a trained team of experts able to compete with the Dutch for new reclamation projects. Here again, the rational Dutch objective is not exclusivity but participation, and their presumption is that the more reclamation that goes on, the greater will be the opportunities for Dutch participation

The Capacity to Collaborate

A distinctive element in the organization of Dutch technical assistance is a quality that characterizes the whole way of life in the Netherlands. This people, tight-packed and vulnerable in a small corner of Europe, buffeted together by the storms of history, have the quality of an extended family and have acquired the habit and the capacity to collaborate to an extraordinary degree. They have learned how to measure each other's capability and to integrate easily into any working task. Against their inability to maintain a large number of specialized scientific or industrial organizations, the Dutch have set that familial capacity for association, for regrouping technical experts and scientific specialists, drawn from the whole range of their private, public, academic, and industrial activities into missions and task forces or teams, to deal with particular problems, including those of AID. That capacity for collaboration, made into an organizing principle within, has been projected outward in the Dutch policy for the international organization of AID. This is surely a capability to be used and a message to be heard beyond national boundaries.

NOTES AND REFERENCES

1. O.E.C.D., *Development Assistance, 1970, and Recent Trends,* pp. 28–30.

2. *Development Assistance Effort and Policies of the Netherlands,* p. 25

3. Ibid., p. 40.

Chapter 8

Science-Based AID
in West Germany

German AID in Its Social Context

GERMAN AID HAS INCREASED SUBSTANTIALLY THROUGH THE 1960S, from $237 million in 1960 to $599 million in 1970, when West Germany stood third in its total AID contribution, behind only the United States and France. German Public AID was .32 of German GNP, as compared to .26 for Japan, .31 for the United States, .37 for the United Kingdom, .63 for the Netherlands, and .65 for France.

Among those variables of the social context that seem particularly relevant in explaining German AID are the following:

(1) the extraordinary vigor and strength of the postwar economy;

(2) the absence of colonial ties. Germany entered the field of action, less, or differently, encumbered by history than France, Britain, and the Netherlands. Unlike them, it was not bound by traditional relationships and responsibilities to particular regions, but rather was free to choose its activities and to engage in those lines of technical assistance that paralleled its interests and its strengths. It inherited no apparatus of empire, seeking to preserve itself. Nor was there an uprooted pool of experts trained to work in low-productivity regions left in the wake of history;

(3) the postwar decentralization of the German political system;

(4) the role of the state in German history, particularly in laying the foundations of industrialization;

(5) the role of the churches as vehicles of moral mission.

Consider the relationship of these to German AID, in policy and practice.

155

The Condition of the Economy

Like Japan, Germany rose out of defeat and destruction to surpass in productivity those who had defeated and destroyed it. By 1970 per capita national income in Germany was greater than any of the other large European countries. Its unemployment rate (0.7) was virtually nil. To fill the growing world demand for its products, it imported thousands of workers from the poor countries of Europe and Asia Minor (Turkey, Spain, Greece) to work in its factories. Exporting far more than it imported, it has accumulated a huge surplus of dollars and gold, constituting a mass of sterile (or sterilized) claims on goods and services that are stuffed into the vaults of the central bank. This trade imbalance, and the correspondingly huge American deficits of dollars and gold, has caused a series of monetary revaluations, in which the worth of the American dollar and of the British pound sterling have declined and declined, in relation to the German mark, as well as to the Japanese yen.

Since real German output is rising so rapidly in relation to German consumption propensities and expectations, and inasmuch as rapidly rising GNP normally produces surplus government revenues, the state is able to spend more for worthy or prestigious purposes, such as AID, without any evident (or felt) increase in the tax burden, hence without provoking political opposition. Given this capacity to spend, it is to be expected that public officials, understandably sensitive as to their status and their national image, would try to catch up to the other European nations and the United States in the level of public AID.

The fact that German industry is overextended, and hence a reasonable objective of German economic policy (as will be seen) is to dampen down demand, the tying of German AID (requiring recipients to confine their use of German loans or grants-in-AID to the purchase of German goods and services) is clearly anomalous. The tying of German AID simply increases the industrial stress and inflationary pressure that German economic policy seeks to reduce. On the other hand, where donors seek to promote exports and to offset underemployment and lagging industrial output (as has generally been the case in Great Britain and the United States during the decade), there is a logic (though it might be short-sighted) in the tying of AID. It is, therefore, not surprising that Britain and the United States are firmly committed to AID-tying. France is equivocal. And Germany champions the universal untying of AID.

> The German government has for years been holding the view that assistance not contractually tied to procurement in the donor country is in the best interest of the developing countries. Therefore, it has at all times adhered to the principle of nontying and, despite the fact that almost all other donor countries are practicing the contrary, actually has refrained as much as possible from tying its bilateral capital aid to German supplies. The proportion of capital aid commitments tied to procurement continued to fall in 1969. Out of a total commitment of DM 1,222.4 million, only DM 399.6 million (33.5 percent) were contractually tied to procurement in Germany. . . . Aid-tying that is still maintained is mainly caused by the fact that the rigid tying practiced by other donor countries is distorting competition and seriously disturbing the normal flow of trade. The German government would be very pleased if the endeavors in the DAC to obtain a gradual relaxation of aid-tying . . . would be successful . . .[1]

> The Federal Government is in principle prepared to reduce aid-tying still existing in capital aid. It will give vigorous support to international efforts directed at a general discontinuance of tying practices . . .[2]

In the logic of national self-interest, the accumulation of investable funds and private savings, at a time when there is a surfeit of domestic demand, plus a critical trade imbalance that has already had grave international repercussions, would lead to other particular AID policies as well. It would lead to a policy of encouraging Germans to satisfy their rising consumption demands by spending more overseas than at home, e.g., through tourism. Thus, the German emphasis, unique so far as I know, on the promotion of tourism in developing countries as a part of AID policy.

> In suitable developing countries, the Federal Government will promote tourism through the improvement of the infrastructure, especially transport conditions, through the building of accommodations and through the training of the necessary personnel, if need be with the help of the development banks . . . through the provision of capital and funds to bolster up investments.[3]

Under these circumstances, national self-interest would also lead to the promotion of German industrial investment abroad and to the extension of German industrial activity and ownership overseas even, under those circumstances, when political risks and rate of return make overseas investment less attractive to the individual capitalists than investment at home. And indeed German AID does lay great stress on the extension of German industrial activity overseas. What is involved in this promotion of investment overseas is so complex and important that it will be dealt with separately in a later section of this chapter.

The Germans diagnosed their economic situation during 1970–71 as one of "overheating" and applied classical anti-inflationary remedies. They raised interest rates, used revenue surpluses to repay the public debt, and limited public spending. The limitation on public spending checked the rise in AID expenditures, and high interest rates burdened the process of economic development. This economic diagnosis and policy prescription are curious. It was presumably the persisting trade imbalance, threatening the whole system of international exchange, that precipitated the German economic crisis. To correct that imbalance would require a relatively higher (not a deflated) price level in Germany, and an increase (not a reduction) in German consumption of imported commodities. If, for other reasons, there was considered to be an excess pressure on German output capacities, then the policy objective might rationally have been to increase the consumption of imports without increasing unduly the demand for domestic production.

The international exchange crisis admits of at least two possible explanations. Comparative market values, i.e., national price levels, may be out of line. In that case, monetary revaluation, hence new exchange rates (bringing price levels into line), might be the appropriate solution. But that need not be (and I do not think, in fact, that it is) the cause of the crisis. In my experience, and in the experience of other travelers, at 1970 exchange rates, the dollar bought more (not less) in the United States than it did in Germany or, I understand, than it did in Japan.

Alternatively, the crux of the problem may be that the consumption propensities in different societies, more specifically (since internationally traded goods and services are a select category, exported-imported and consumed at manipulated prices and under highly constricted conditions) import consumption propensities, are out of kilter.

Supposing that that is the case, e.g., that with import prices rigged as they are, and with income distributed as it is, and with tastes and spending habits given, the individual German consumer will not buy on the market and consume as large an amount of imported goods and services as he could and (for the sake of trade equilibrium) should—then what is to be done? Change the constraints and restrictions upon imports? Redistribute income? All these have repercussions that the German government is disinclined to risk.

An extremely simple alternative is entirely neglected by the Germans, the Japanese, the French, and, in the days of its own trade surpluses, by the Americans and by every other Western country. The multi-billions of dollars and of gold piled in the vaults of central banks as a consequence of trade disequilibrium could be spend directly by governments in possession of those surpluses, on public account to procure goods and services from other countries, not for the purpose of war but for the welfare of their own people *or for AID* to those in need. The import-consumption propensities of the public sector, which are infinitely more subject to deliberated manipulation and precise control than the import-consumption propensities of the private sector, ought also to be brought within the scope of international trade policy with AID expenditures used to acheive a trade balance.

The Promotion of Overseas Investment

It has been in the liberal tradition for governments to prefer, and to favor, investment overseas in contrast to investment at home. The reason for this preference is obscure. In British experience, which has profoundly influenced liberal ideology, centuries of investment overseas did provide an income base for the bountiful lives of the leisure class. Income from abroad, moreover, gave, and still gives, to its recipient a degree of autonomy vis a vis the political authority of his own country, and a privileged position in avoiding the bite of taxation. Yet whatever its benefits to the leisure class, inasmuch as investment had an effect upon productivity, the diversion of investment overseas contributed to the decline of the British economy. The lesson was lost upon the Americans. In postwar decades American tax policy has greatly favored, and continues to favor, investment overseas in contrast to investment at home; and the extension of corporate investment and activity abroad has been accompanied by a slowdown in growth and a relative decline in the domestic economy of the United States.

In Germany the situation seems to be somewhat different. Whatever its national aspirations, Germany has never been a colonial power. The bourgeois outlook in Germany has not been oriented to investment opportunities overseas, and the focus of investment and of business endeavor has, perforce, been European. The Germans now find themselves with a surplus of claims on foreign goods and services, greater than the propensity of German consumers to consume and beyond the initiative of German business to invest. Moreover, the world demand for German products has been beyond their capacity to produce with the available labor force. In order to meet that demand, they have imported a great number of workers from the poor countries of Europe and Asia Minor. Consider what might be said concerning the German response to this situation, and concerning their present policy options.

(1) The Germans need not have brought in foreign labor. They could have adjusted to excess demand by increasing export prices or placed a greater emphasis on expansion through more advanced technology. In that case, it could be argued, real income per capita would have increased more rapidly than it has, but the relative scarcity of labor might have forced a redistribution of income in favor of the wage earners and a redistribution of power in favor of the trade unions.

(2) Alternatively, the German policy makers could have reasoned, "What we are selling to the world, and what the world values is our organizing and high technological capabilities—which we have in abundance, though we may have come to the limit of the available, ordinary, factory-trainable labor. By bringing in overseas labor, we are making the most of (and optimizing the marketable values of) our peculiar German capabilities and, hence, optimizing the output contribution of, and the income flow into German society." On the basis of that rationale, perhaps, migrants were brought en masse to work in German factories.

(3) Presumably, the migrants have benefited, for they have been trained, they have learned, they have earned far more than they could have at home. Presumably, the Germans have benefited by being able to utilize more fully and increase the real return on their particular capabilities and skills. On the other hand, to bring in a massive minority—different from the majority in appearance, culture, manners and customs, education and skill, religion and self-identity, and destined for a subordinate position with respect to the majority—bodes grave problems, tensions, and difficulties for both sides in the future.

(4) Some Germans in government and in business see another option, one that would permit them to exploit the values of their higher technology and organizing capability in order to increase German production without importing foreign workers. The policy maker reasons as follows: "We have learned how to train and organize labor echelons of Greeks, Spaniards, and Turks to work effectively in new factories in Germany. We could as well organize them to work in new factories in Greece, Spain, and Turkey, producing components that would be fed back into the processes and the final outputs of German industry. That way, the poor countries could benefit. The workers would not be uprooted. Germany would be spared the threat of intergroup tension and trouble. And besides, in building those factories and in preparing the required infrastructure, our foreign exchange surplus, now a wasting resource, can be put to good use."

This option seems now to have captured the imagination of German authorities and to have significantly shaped plans for the transformation of German AID.

(5) To achieve the implicit objectives of that option would require an *integration* of industrial planning and of AID programs. It would require that the government provide sufficient incentives to offset political risks and the inhibitions of private business to venture into unfamiliar territories. It would require the development through capital AID or otherwise of an infrastructure (transportation, energy and water supply, executive housing) appropriate to the planned industrial operations. It would require the training of workers and managers abroad, through AID programs coordinated with the building of factories and the opening of employment opportunities for them.

(6) The German government has undertaken to support and to promote private investment abroad (and not only to developing countries) by offering tax advantages for such investment, guarantees and warranties on exports, information on overseas investment opportunities, through the negotiation of "investment promotion agreements" with various countries, and through low-interest loans "to enable enterprises willing to invest in foreign countries to establish subsidiaries." Through 1969 DM 127.3 million was committed by the government to that purpose.

It must be insisted that the aforementioned is not intended as a description of practice and policy through 1971, but of an emerging idea that has captured the enthusiasm of some German officials, which is interesting of itself and which has behind it the logic of national self-interest and that will be made manifest, I judge, in future AID programming.

Any AID policy or program that ties directly into the economic self-interest of the donor becomes ipso facto suspect in many eyes. Yet, surely, support in the processes of economic development will find its most solid foundation where it is directly and unequivocally based on mutual benefit rather than on charity. Yet charity too is a powerful force for AID, as expressed in Germany through its churches.

The Churches and German AID

As in the Netherlands, the churches are the single organized force in support of AID programs. Moreover, the churches actively engage in development projects in cooperation with the government and also on their own. Thus, in 1969:

The government supported the implementation of 100 important Church projects in the developing countries by providing Federal budget funds amounting to DM 67 million. Federal budget funds totaling DM 61.1 million were disbursed for newly approved projects already under way. An important precondition for the promotion of a Church project with federal funds is that the sponsoring agency must contribute at least 25 percent of the total costs of the project. Measures were concentrated mainly on education, public health and social welfare. Special emphasis was laid on the setting-up and expansion of secondary schools, teacher training institutes and universities, hospitals and training facilities as well as social centers and centers for adult education. To an increasing extent, federal funds were set aside for projects in the industrial and agricultural sector (such as training workshops, handicraft schools, agricultural training centers, auxiliary and extension services for cooperatives) and Community Development projects .[4]

The two Christian Churches continued, in 1969, the donation campaigns "Misereor" and "Brot für die Welt" which they began ten years ago. From the proceeds which amounted to about DM 82 million a large number of development projects primarily in the educational field, health and social assistance were supported. For the first time in 1969, funds from Church tax revenue were made available for development projects. Following decisions taken by both the Roman Catholic and the Evangelical Church, these funds are to be increased considerably in the coming years and to attain up to 5 percent of the total Church tax revenue.[5]

Vocational Training

Industrialization in Germany did not emerge spontaneously—did not evolve organically out of the initiatives and innovations made—out of the skills acquired and the lessons learned by a myriad of individuals in pursuit of private gain in the marketplace, as was the case in England, the United States, and France. In Germany, industrialization was a child of the state. Industrialization in Germany was the precursor of industrialization in Japan, Communist Russia, and China, and of all the efforts to industrialize associated with AID. For in each of these, whether through centralized planning or through the deliberated effort to harness and promote individualized self-seeking, industrialization or the effort to industrialize was through the policy and volition of the political authority.

To provide the foundations of industrialization, the German state organized the first universal system of education; and this system was intended to inculcate the peasant mass not only with literacy, but with the practical knowledge and operating skill required at every level and in every aspect of the modern economy. That emphasis remains. No country, to my knowledge, has developed so fully and seriously its system of vocational training as has Germany.

The German capability for vocational training responds to a fundamental need of the LDCs—namely, to build into their economies the requisite cognitions of mechanism, technique, and process. Correspondingly, the bulk of German technical assistance (in 1969 approximately 75 percent) is related to the tasks of vocational training.

In Germany, also, the state first mobilized science resources systematically in support of technological advance in industry. Numerous institutes for industrial or agricultural R and D remain, and the integration of some of these into AID programs came about in a rather particular way.

The Decentralization of Political Power and the Federal Institutes

Since World War II, political control in Germany has been decentralized among the states (or *Lander*). This has greatly reduced the function and responsibilities of some of these research-based instruments of collective action—e.g., in agriculture, industry, geology, fisheries, forestry—created to serve the former centralized regimes. Some of these long-established, science-based institutions—depleted of functions by the decentralization of German political power, and in search of a new role and functional outlet —became at least temporarily available as instruments for technical assistance.

Four of the federal institutes engaged in AID programs are concerned with geology, forestry, fisheries, and metrology. These will be considered in turn.

Geology

The primary focus of the Federal Geological Survey or Bundesanstalt für Bodenforschung in Hannover is on mining and geology, but its interests encompass all subsoil studies—hydrological, pedological, geophysical, paleontological. Domestic geological surveys and services in Germany are no longer performed by the Federal Agency but by geological organizations attached to the separate *Lander*. Hence, operational responsibilities open to the Federal Survey are confined to the external interests of the republic. For that reason, its present task has largely to do with technical assistance programs to developing countries. In 1969 the Federal Geological Survey employed 154 scientists, about half of whom were engaged in foreign projects. The heads of the geological agency also hold professorial chairs in German universities and, thus, are actively engaged in educating, selecting, and recruiting their research personnel.

From 1963 to 1969 the scientific personnel attached to the institute has increased by 50 percent, but expenditures on technical assistance projects increased from DM 1 million in 1963 to DM 7 million in 1970. A substantial but diminishing proportion (18 percent in 1970) of the Institute's work overseas has been on private account.[6]

The Institute works closely with UN agencies, with ECAFE and, depending on the political climate, with French teams of the B.R.G.M. It has carried on mineral explorations in every corner of the globe. Increasingly, as its involvement in technical assistance programs has increased, its efforts have focused more on the search for underground water supplies and on the study of soil structures in relation to the development of power sites, waterways, irrigation projects.

It does cartographic studies for UNESCO. It maintains permanent missions for research and teaching in a number of countries, and is continuously engaged in counterpart training.

Forestry

German science has a traditional interest in world forestry. It was mostly the Germans who started the disciplined study of forestry and the scientific control of forests throughout the world, including, for example, in the United States and in the British Empire and Commonwealth. The science of forestry and techniques of forestry management are of particular importance to developing economies, in the conservation of their soil, as a basis of managing their lumber resources, and for the establishment of wood-based industries. Of recent years, the Russians have led in developing the means of establishing wood plantations in semiarid areas.

German forestry institutes are divided between those attached to the *Lander,* servicing the domestic forests and forestry-product interest and industries, and the Federal Institute for Forestry and Forest Products in Reinbeck, near Hamburg. The Federal Institute does research on problems of general (or "basic") interest to German forestry, but its tie-in to operations is entirely overseas. In 1971 it employed eighty scientists, double the number employed in 1963. The Federal Institute services its ministry in the evaluation of forestry-related AID projects. It maintains a documentation center and reference service for foresters abroad. Along with *Lander* institutes, it supplies experts for international missions. Concurrently with the University of Hamburg, it acts as a specialist training center. Finally, it provides the headquarters base and the logistical support center for a number of continuing AID projects. These projects have included the establishment of institutes for forestry engineering, agencies for forestry surveys and inventories, and university departments for the study of forestry and soils. Currently (1971), research emphasis is on mechanization, cost/benefit and feasibility studies, and the rationalization of forestry operations.

Fisheries

Fisheries research in Germany is highly evolved, and its complex of research institutes are among the strongest in the world. The *Lander* have their institutes of inland fisheries. At the Federal government's Bundesorschung-sanstalt für Fischerei in Hamburg, there are separate institutes specializing in the problems of: (1) storage and preparation, (2) high-seas fishing, (3) coastal fishing, and (4) fishing techniques. The latter institute is also concerned with the technology of shipbuilding and works with producers of fishing equipment and gear. As a part of this complex of institutes are schools for training fishermen, and, at the pure science end of the spectrum, the famous station for research on marine biology at Helgoland. Sea mapping and oceanography are under the Ministry of Transport.

Clearly, this research and developmental competence in fisheries is relevant to the German capability to assist in economic development abroad, and a part of German AID has been related to the development of fisheries.[7] An example is in the introduction into Thailand of a trawling technique which, between 1960–71, is said to have increased the catch there from 221,000 to 1,000,000 tons of fish per annum.

According to Von Brandt, who headed the Institut für Fangtechnique (fishing technique) at the time that the Germans first introduced their trawling technique to the Thais, the Japanese had tried to teach their technique for fish trawling just before the Germans came, and had failed. The German technique was not intrinsically better, but better adapted to the situation in Thailand. The Japanese method requires simultaneous action by two cutters, and skills which are evidently more difficult to impart and acquire. Moreover, the use of two cutters working in tandem goes against the institutional grain in Thailand, where most boats are individually owned. This illustrates how advanced techniques may be more or less well suited to the circumstances of particular low-productivity societies; it suggests, indeed, that the optimum use of advanced technology would require the capacity to develop and adapt, continuously and as a dynamic response to experience what is best suited to the particular locale, hence the need for R and D to shape and improve techniques in terms of LDC need and circumstance.

Fishing operations in Africa have been established through the work of the Institute; in Lomé as part of the development of a viable seaport, and in Togo where two modern deep-sea cutters were supplied along with facilities for net making, fish preservation and fish distribution, and where men were trained for every level of operation.

Metrology

The Physikalisch-Technische Bundesanstalt (PTB) in Braunschweig and Berlin was founded in 1887 to "foster by experimental work the exact sciences and precision techniques." It is the central West German agency for the development of standards of measurement and quality norms, and for research and analysis in the measurement of physical phenomena, in relation to theory, technique, and applications. It has major research departments in: (1) mechanics, (2) electricity, (3) heat, (4) optics, (5) acoustics, (6) atomic physics, (7) engineering. In 1969 it employed 1,250, of whom 300 were in the scientific grades, and it had an annual budget of about DM 40 million. The Institute provides a vital component of the science infrastructure in Germany.

In 1958 the PTB undertook, as a German AID project, to establish at a cost of DM 50 million an institute for material testing and for measurement (metrological) research in Egypt. By 1970 that effort had run afoul of political stress and had been abandoned. Nevertheless, the German effort and experience raises a number of issues and suggests problems and possibilities of general interest.

The objective was to build into the Egyptian economy those testing and measurement services required as support for the modernization of industry, particularly in textiles, petroleum, coal, and other mining industries, thus to create, with the limited means available, basic components of a science infrastructure. The required services were conceived by the German planners of the PTB as a three-level activity: Level C, Level B, and Level A. At the C-Level, would be the private or public agencies engaged in the practical testing and measurement in industrial or governmental choice and operations. The C-Level would include the application of the quality controls, measurement of standardized outputs, systematic evaluations of alternatives available for procurement, prototype testing in preparation for the output of new products, in sanitary inspection, in the enforcement of fire protection and of building codes, and in test of product or equipment alternatives prior to public procurement. C-Level activity would be carried on in engineering operations, industrial laboratories, commercial laboratories, government sanitation bureaus, pest control services, agricultural stations, building-inspection services, and so forth. In the view of the German planners, the ultimate objective of the project would be to establish a viable C-Level activity and progressively to raise the effectiveness of that activity. Only at the C-Level could there be a payoff in social benefits.

C-Level activity needed technical personnel and institutional mechanisms; the former had to be trained, and the latter often created anew. The

Germans found, for example, that standards and quality controls enforced by private industrial associations in Europe would have to be incorporated into the law and enforced by government inspection in Egypt. Formulating and promoting the enactment of such an appropriate legal code became one of their tasks. Higher B and A level research is justified if, and to the degree that, it is required in order to increase the effectiveness of services at the C-Level.

At the B-Level would be formulated general codes and standards adapted to the particular problems and needs of the economy. Through B-Level studies, and through a B-Level search of published data, information which is relevant and useful to materials evaluation, testing and measurement needs and practices in Egypt would be systematically gathered, organized, and made available for C-Level activity. B-Level research, also, would provide the complex equipment and specialized skills needed to resolve those problems of practical measurement and testing which are beyond routine analysis by the commercial laboratory and are outside the ordinary engineering competence. B-Level research would constantly examine world developments in the sphere of measurement and testing to determine their relevance to Egyptian practices and needs. It would seek to adapt new techniques, mechanisms, theories, and practices in materials testing and metrology to the circumstances and needs of the Egyptian government and Egyptian industry. It would inform government and industry of the values of new and improved standards and methods, and would instruct those performing C-Level services in the application of new and improved techniques and mechanisms.

Finally, A-Level activity would consist of exploratory scientific research of a fundamental character, e.g., the study and measurement of surface tensions, which, indeed, might result in a deeper understanding of relevant physical phenomena, but with no *direct* relation to specific services and concrete problems at the C-Level.

The Germans proposed to start by creating the appropriate B and C Level institutions. But the Egyptians saw the matter differently. They demanded an A-Level institute that would perform the highest, most abstract, purest, most fundamental research, at once. After all, Egyptians also must share in the glories of High Science. They could not be satisfied with crumbs.

The German experts refused to start with an A-Level institute. This, they considered, would not be admissible in view of the needs of the Egyptian economy, the lack of Egyptian scientists, and the scarcity of the German scientists, who would be obliged to man operations until a sufficient number of Egyptians were trained. They proposed rather to create several B-Level institutes, one at a time, coordinating these with the development of C-Level activities. They proposed that A-Level research should be allowed to develop in due course out of the problems and experience generated at the B-Level. Ostensibly, the Egyptians accepted and agreed to build the facilities, while the Germans would furnish equipment and the scientific personnel for a limited number of years, and, also, would train counterpart Egyptian scientists in Germany eventually to relieve the Germans sent to Egypt.

Meanwhile, the Egyptians entered into negotiations with UNESCO, with the Germans unaware of those negotiations and with UNESCO unaware of the agreement with Germany. An agreement with UNESCO was concluded whereby that agency undertook to establish an A-Level institute in Egypt. While this diplomatic coup was a display of negotiating skill and diplomatic adroitness on the part of the Egyptians, it hardly represented the optimal method of organizing technical AID, particularly with respect to the use of science resources. It would, indeed, explain why donors need a coordinated approach in planning technical assistance. In any case, the eventual organization of the A-Level institute, to be financed by UNESCO, was also left to the Germans.

There was no smooth sailing. Egyptian ministerial responsibilities were changed and changed again. Eventually the project was abandoned as a consequence of diplomatic tensions between Germany and Egypt.

The numerous requests that the PTB establish similar institutes in other countries raise the question as to how best to build this essential component of the science infrastructure into the economies of the developing world. On rational grounds, the appropriate organization at the C-Level would be local; at the B-Level, regional (serving common-language outlets, and areas with similar industrial-agricultural activities at an analogous stage of economic development); and at the A-Level, international and supranational.

In 1971 AID arrangements with the PTB to establish or upgrade the metrological capabilities were in effect in Brazil and Argentina, and a similar arrangement with India was being prepared.

The German experience illustrates very clearly the potential values of donor R and D organizations as an instrument of AID. In the German case, however, that potential has barely been realized. The Federal Institutes have only been modestly employed in AID activities. Nor have they been given the opportunity to initiate AID programs or projects, nor to develop their capabilities in relation to AID objectives.

Institutes, Foundations, Consultants

German AID has benefited from a strong interest on the part of intellectuals in economic development, reflecting perhaps a restless energy of the young who look outward from and seek outlets beyond the nation. In any case, there has been an extraordinary proliferation of public and privately supported institutes, more or less autonomous, where team research is carried on, American-style. Thus, in 1963 there were at least 130 different institutes, associations, or foundations, mostly in the social sciences, concentrating on problems of economic development.[8]

A word concerning them is in order.

Research in the regular departments of a German university is normally dominated by the sovereign interests of a single professor. Whatever the values of this system, it does not lend itself to continuous team research in selected areas of public concern. Independent research institutes have proliferated in part as an escape from professorial power.

Some of the development-oriented institutes consist of a man and a desk. Others are large and well-endowed. The German Foundation is an information-documentation center that sponsors conferences and seminars on a variety of development-related projects.

The German Research Association in Bad Godesberg, a quasi-public body, partially combining the functions of the National Science Foundation and the National Academy of Sciences, has at present only a peripheral role in the organization of science resource for technical assistance. It advises the government on experts who might be available for special assignments. It has been concerned with harmonizing the interests of university scholars in particular fields of development-related research and the interests of public agencies in the implementation of AID policies.

Following the pattern of the Ford and Rockefeller foundations in the United States, there are also German (Thyssen and Volkswagen) foundations that support various scientific activities. The Thyssen Foundation has sponsored substantial development-oriented projects in Kenya and Tanganyika.

Private Consultants

The Society of Consulting Engineers in the Federal Republic of Germany lists almost a hundred private consulting firms, including a broad range of scientific skills and technical expertise. The use of consulting engineers in Germany, following United States and British practice, was wholly a postwar development. This new German consulting industry is dependent on the world market for technical assistance. Domestically, German consulting firms have faced the hard antagonism of large industrial companies, who follow a policy of technical autarchy, and look upon the newcomers as a threat to the fealty of their own expert technical personnel. The consulting fraternity provides the German government with an expert, objective instrument of evaluation, knowledgeable with respect to German scientific and industrial capabilities as well as developmental needs, to analyze and judge project proposals from developing countries and to prepare feasibility studies.

Planning and Programming Technical Assistance

Prior to 1971 German technical assistance has been in the form of ad hoc projects. These "projects" characteristically originated with requests from LDC recipients, submitted to the German embassies and forwarded to the Ministry of Foreign Affairs in Bonn. The Ministry of Foreign Affairs sent them as project applications to the Ministry of Economic Cooperation (BMZ), which controls the allocation of the government funds appropriated for technical assistance and also to the particular ministry with functional responsibilities related to the requested project and having command of the requisite technical resources, e.g., the ministries of Agriculture, Industry, Commerce, or Education.

These ministries send the application to experts within or outside the government—e.g., to engineering and technical consultants, scientific institutes, or university professors—for an independent evaluation. In making that evaluation, experts might travel to the country making the request for an on-the-spot study of the feasibility and possible values of the project. The evaluated application works its way back, through the levels of bureaucracy, eventually to be approved or disapproved. If approved, it is implemented through the action of various ministries.

The system has had a number of conspicuous defects. It divides and subdivides authority without a functional necessity for such division. It passes decisions from hand to hand along long bureaucratic corridors. It puts the power to initiate in the hands of recipients, who can hardly preconceive or judge the resources that are, in fact, available. It offers no evident opportunity for scientists and research institutions in Germany to initiate projects or to participate in development planning. Nowhere is there any place for a national balancing of available resources against the comparative needs and values of AID alternatives. Projects are so heterogeneous and scattered that it is difficult to determine their value. Indeed, whatever their intrinsic excellence, they may well be quite valueless: a fine road that leads nowhere, a true resource inventory that is never used, men trained in skills for which there is no employment opportunity.

Change is in the offing. It is the announced intention of the government to put aside the system of ad hoc projects in favor of complex AID programs that would gear into the plans of the recipient country's development authority. Such programs would integrate the contributions of capital AID, technical expertise from public agencies, social planners, research institutes, the churches, and private enterprise. Hopefully, the programs or contributions of various national donors will be coordinated through an international agency. Thus, a "Transition to long-term, integrated, country-related and internationally coordinated aid programmes" is proposed as the first of the AID policies of the Federal Ministry for Economic Cooperation.

> The instruments of German development assistance in the future will be most effectively applied to long-term, integrated, country-related and internationally coordinated aid programmes for individual developing countries . . . drawn up in a dialogue . . . with the developing countries . . . and in coordination with multilateral and other donors.
>
> These aid programmes . . . will permit not only the examination of the projects in individual cases as to their economic and social effectiveness, but also the selection of priorities, locations and combination of projects within the framework of sectoral, intersectoral or regional interrelationsships. . . .
>
> The aid programmes will constantly be brought up to date and adjusted to changed conditions in the developing countries, through overall and sectoral analyses as well as on the basis of observation and inspection of the projects. . . . By means of the aid programmes, the interplay of all the instruments of

public development aid will be achieved. The programmes will include credit aid and grants, make possible the coordinated participation of private sponsors and volunteers of the German Volunteer Service (DED), and provide appropriate pointers to private industry . . .

The criteria for this selection are, among other things, the stage of development of the beneficiary country, its capacity to absorb foreign capital and know-how, its development prospects, as well as its own efforts to attain its development objectives.[9]

NOTES AND REFERENCES

1. Memorandum submitted by the Federal Republic of Germany for the DAC Annual Aid Review of 1970. *German Development Assistance Policies of 1970* (Bonn: Press and Information Office, 1970), pp. 22–23.

2. *Development Policy Concept of the Federal Republic of Germany for the Second Development Decade* (Bonn: Press and Information Office, 1971), pp. 18–19.

3. Ibid., p. 14.

4. German Development Assistance Policies, pp. 33–34.

5. Ibid., pp. 40–41.

6. See Bundensanstalt für Bodenforschung, *Tatigkeitsbericht 1969 und 1970,* (Hannover, 1971), pp. 3, 4.

7. Cf. W. Becker "Die Bedeutung der Fischwirtschaft für die Volkswirtschaft der Entwicklungslander und deren Forderung durch die Bundesrepublik Deutschland" *Berichte Über Landwirtschaft* (Hamburg and Berlin: Verlag Paul Parey, 1964).

8. Cf. Deutsche Stiftung für *Entwicklungslander in Forschung und Lehre, 1959– 1962* (Berlin-Tegel 1963).

9. *Development Policy Concept,* pp. 10–11.

Chapter 9

Science at Sea

An Alternative Approach

HITHERTO WE HAVE APPROACHED THE ORGANIZATION OF SCIENCE resources for economic development as an aspect of national policy, analyzing the activities, impacts, and problems of public agencies. The matter could have been dealt with differently. Scientific activity, transcending the politics of nations and the agencies of government could have been our starting point, with that activity studied for its actual and potential effects on the specifics of development. In this final chapter we shall, tentatively and speculatively, look at the question of mobilizing science for economic development from this alternative point of view.

What is intended is neither a definitive analysis nor a factual survey, but rather an exploration and speculation upon an alternative approach. What field of development might we select? Certainly there are many that might reasonably be studied. Here we have chosen to consider the development of fisheries and the science and research related to it.

This chapter will be divided into three parts. The first presents a critique of certain commonplace outlooks on science, in relation to technological advances. The second proposes, in my view, a reasonable way of conceiving science in relation to the economy. The third analyzes the relation of science to the world development of fisheries.

PART I. FALSE IMAGES

The Scientist as Magician and Gadgeteer

A popular image of the scientist, somewhat shopworn now, has been that of a man in white coat, armed with Bunsen burner and test tube, with trusty slide rule, computer, or atom smasher and nuclear reactor, at his side, making discoveries that turn the world topsy-turvy, like a magician pulling rabbits out of a hat. In a flash, he produces marvelous thingamajigs that open the way to a better life for all humanity, and especially for the teeming starving millions in the underdeveloped lands. Journalists have proclaimed science's role in economic development in wide-eyed announcement of new marvels, miracles, promises. Consider this one. The *New York Times* on December 26, 1965, ran this front-page headline.

FISH FLOUR TO FEED HUNGRY PERFECTED

It was a Christmas gift and benediction to the world: "a small group of Federal scientists working quietly in laboratories and in a pilot processing plant in nearby Maryland" seemed to have produced a major breakthrough that would end protein starvation, reduce disease, raise productivity for two-thirds of the human race. A miracle, a marvel, a wonder, certified by no less an authority than the National Academy of Sciences and the National Research Council. The story read:

> Federal scientists have made a major breakthrough in the universal search for a solution to the world food crisis.
>
> The answer, already perfected by experts of the Interior Department's Bureau of Commercial Fisheries, is a new process for producing a clinically pure fish concentrate with a rated protein content of 80 percent. Eventually the concentrate could end "protein starvation" for about two-thirds of the human race.
>
> Feasibility studies indicate that if only the unharvested fish in United States coastal waters were translated into the concentrate it would provide the normal protein requirements for one billion persons for 300 days at a base production cost of half a cent a person a day.
>
> Government officials consider development of the new fish concentrate to be dramatically significant in terms of the mathematics of increasing populations, declining food production and malnutrition in the underdeveloped countries of Asia, Africa, and Latin America.
>
> Studies by the United Nations Food and Agriculture Organization and World Health Organization have pinpointed protein hunger as the most

pressing human problem of the century. Over 80 percent of the world's population does not receive sufficient daily protein, while some 60 percent of the world population verges on actual protein starvation.

A recent task force report to President Johnson brought out that about 50 percent of infants and pre-school-age children in underdeveloped countries suffered from protein malnutrition. There is evidence that the condition retards mental and physical development of between 10 and 25 percent.

The task force report said that overcoming the vitamin and protein deficiencies of young children in most underdeveloped countries would do more to reduce disease and eventually raise productivity than any other health measure that could be taken.

At last we could breathe easily with respect to those unfortunates who suffer and starve in distant lands and shores so far away.

But certain doubts arose. More than three years earlier, the selfsame agency had produced a fish flour for which exactly the same claims were made. What had happened then? The United States Food and Drug Administration had forbade domestic sales for human consumption on the purely esthetic grounds that the whole fish, including head, tail, fins, and viscera were used. The great BREAKTHROUGH then had not produced a cheaper, more palatable, or more nutritious fish flour—it had succeeded simply in finding a way around the interdiction on domestic sales in the United States. Why was this important for the salvation of the teeming millions in far-off lands? After all, the FDA had had no objections to foreign sales or giveaways. The article explained:

> Government officials promptly dismissed any such course of action [as giving the fish flour to the LDCs] on grounds that it risked a Communist propaganda attack charging the United States with palming off on others a product it considered unfit for human consumption at home.

Those public officials must have been a cold-hearted lot to refuse to save teeming millions (two-thirds of the human race) in Latin America, Asia, Africa from (no less) protein starvation simply because the bad old commies might score a propaganda point! And anyhow, with a discovery so marvelous, with benefits so great, why had not the other donors, the French, the British, the Japanese, even the Russians themselves, used it in their programs? Indeed, why had not the developing countries (who have organized far more complex operations than fish flour factories) followed this quick and easy plan to salvation?

I remembered that what was being said about fish flour had been said years earlier about the soybean, soya flour, soy sauce, as a food additive and a cheap protein source, and that many generations ago Victorian Englishmen had proposed, with the same enthusiasm that the water used to cook their vegetables should be saved and used again to provide an adequate diet for the working classes. And, besides, surely, to harvest all the unharvested fish in American coastal waters is no casual and costless affair, but a massive, expensive enterprise with its impact on future fish populations. And in fact that miracle announced by the *New York Times* on December 26, 1965, was lost in the limbo of forgotten things by December 26, 1966.

Fish flour is possibly a useful food additive and, with the government as the buyer, a profitable sales outlet for inedible fish. But "a major breakthrough in the universal search for a solution to the world food crisis!" Think again!

There have been numerous grand breakthroughs, allegedly opening new vistas for development, ushering in new epochs; satellites carrying television to every remote village, solar-energy machines of every sort, atomic explosions to drain the swamplands of the Amazon basin, nuclear reactors for atomic energy, desalinization machines to make ocean waters drinkable. The last two especially have been greatly publicized, and have been heavily invested in by donors and by developing countries. However, none of these schemes has made any significant contribution to economic development. Consider atomic energy and desalinization.

An atomic reactor is an expensive, complex apparatus for boiling water. Steam from boiling water, no matter how that water is boiled, can, through conventional means, be turned into electrical power. As a system for boiling water, the nuclear reactor has a certain advantage in areas remote from regular fuel supplies or otherwise difficult of access, and hence has a certain relevance to development planning in some low-productivity economies. But no system for boiling water is an open sesame to economic development; there are many countries where cheap fuels are directly available and are daily being pumped in enormous quantities from the earth (Malaysia, Arabia, Venezuela) or where great hydroelectrical power sources remain untapped, yet which remain technically backward and with a very low productivity. The enthusiasm among the seers and prophets of science and among government officials for the developmental potential of nuclear reactors as systems for boiling water among the impoverished nations cannot be accounted for on rational grounds.

Desalinization techniques have been developed which can produce fresh water from the seas, but at a price. The price will be lower where the cost of power is less and where the complex mechanical facilities can be set up, operated and maintained efficiently and cheaply. The net value of the operation will be the difference between the costs of producing the incremental fresh water and the benefits to be derived from its use. Even if fresh water could be produced from the seas without costs, it would be no guarantor of growth. After all, water is a free element in many low-productivity countries which nevertheless remain poor and technically backward.

The public's memory is short; it has never learned to measure new promise against old performance. So long as the people look for miracles, journalists need headlines, and officials want easy ways out, the magicians of science will pull from their hats breakthroughs to economic salvation.

It is not our purpose to denigrate any technological innovation or adaptation. Fish flour, nuclear energy, desalinization, and the others have a part to play—along with the diesel engine, the typewriter, the photo-offset printer, the bulldozer, the tractor, the telephone line, the Bessemer converter, the thermocouple, the computer, radioisotopes,—among the techniques that might be incorporated into the economies of developing countries. What is wrong are the claims and the sort of expectation that brings forth such claims. All are bits and pieces, in a virtually infinite universe of technical potentialities, which must be adapted, fitted together, set in motion as a complex and effective process of producing and distributing goods and services. If there existed among all societies a capacity for finding, adapting, fitting together, and seing in motion the range of technologies already known and practiced, then there could be no problem of economic development, no problem of hunger and poverty for the teeming, starving millions in Asia, India, Africa, Latin America. But this capacity does not exist over a great part of the world; and where it does not exist, then no increment to the universe of technologies can lead to development. The need is not for specific technological breakthroughs or scientific miracles, but for a capacity among the impoverished peoples for coherent social action, including action in organizing the science resource and using the accumulated information that science offers.

Science-Input, Progress-Output

Economists are habituated to considering all economic processes as a series of input-output relationships. This is "commonsensical," and often useful in explaining industrial production. So much input of coal, coke, limestone, iron ore, truck-time, machine-time, labor-time results in an output of so many tons of steel. And when the economist evaluates a particular course of action, he tries to measure and compare the cost of the inputs against the values of the outputs.

But if the use of science resources is considered as an input, what is the output? If science is cause, what is effect? Economists, more uncertain today, not long ago would have answered in a single voice "technical progress." One Establishment group wrote a book in which they called R and D an "inventions industry." Put more money (resources) into R and D and out comes more inventions, equals more innovations, equals more technological advances, equals higher productivity, equals a greater GNP and more for all. Does this cause-and-effect relationship actually exist, or will this approach lead to delusion (and eventually to disillusion)?

Since our concern is with the relationship of science to the development of fisheries, it is of some interest to consider as an example of this science-as-an-input, technical-progress-as-an-output paradigm, in a study conducted in 1964 by the National Academy of Sciences and the National Research Council purporting to analyze the costs of research in oceanography[1] in respect to economic benefits to (and, consequently, economic growth in) the United States.

According to this study,

Federal support of oceanographic research has increased substantially from $24 million in fiscal year 1958 to $124 million in 1963. According to tabulations of the Interagency Committee on Oceanography of the Federal Council for Science and Technology (ICO), the federal budget for oceanographic research in fiscal year 1965 is $138 million. Projected budgets indicate an annual growth of about 10 to 11 percent, reaching $350 million in fiscal year 1972.[2]

In the light of these projected expenditures, the National Academy "attempts to estimate some of the future economic benefits that could result from oceanographic research and to compare them with the costs of doing the research."[3] Various benefits are imputed,[4] but here only those accruing from United States fisheries production will be considered.

The report finds that average annual investment of $50 million per year in oceanographic research would yield future benefits in "increased annual new production" of $555 million per year, to be realized in five to fifteen years. Using *discounted*[5] values, the report calculates the benefit of oceanographic research to be 5.8 times greater than costs.

With regard to the economy of marine fisheries, the report has the following to say:

> The total annual production of the world's marine fisheries increased from 25 to 40 million metric tons between 1955 and 1962. . . . The world's industrial uses of fish are increasing more rapidly than the use of fish directly as food for humans. The latter, however, still shows an average growth rate of 5½ percent per year, more than twice the rate of growth of the world's population.
>
> Most of this increase in fish harvest is a result of the activities of other countries [than the U.S.].
>
> In the United States. . . . The total supply of fish destined for direct human consumption was 4,020 million pounds in 1949 and has risen to 4,593 million pounds by 1962. During these 14 years the share caught domestically declined from 3,305 million pounds to 2,523 million pounds, while imports increased from 715 million pounds to 2,070 million pounds. . . .
>
> A large share of this imported food [and fishmeal] is produced by companies owned in whole or in part by U.S. operators.[6]

Given this economy of marine fisheries—what sort of benefits (and in what magnitude) will accrue to the United States as a consequence of oceanographic research? The report answers:

> Increasing the U.S. domestic catch of fish requires the existence of sufficient additional productive potential of fish stocks accessible to our fishermen, and the existence of markets for the catch. Both of these conditions we believe can be satisfied if the necessary research is done.
>
> . . . if our fishermen, through research and engineering, can recapture the share of the market lost to imports during the past decade and a half by cutting their production costs, an annual market for nearly 800,000 tons of

edible fish and a similar amount of industrial fish would be provided. Additional markets exist in other countries if prices are competitive. . . .
. . . Rational development of the U.S. domestic fisheries could result in doubling U.S. production in 15 years. The growth of overseas fisheries of the United States . . . increasing by a factor of four within a decade.[7]

In essence, the "rational development of U.S. domestic fisheries," will permit American fisheries to recapture a larger part of domestic and world sale of fish. But why is oceanographic research required for or why should it lead to the "rational development of U.S. domestic fisheries"? According to the report,

(1) For those fish populations being substantially exploited it can provide the basis of more efficient catching operations and the basis of "conservation" (maintaining the populations at levels that will produce maximum yields year after year). . . . (2) For the populations that are little used, or not used at all, research on their habits and reactions to the changing sea can provide the basis for developing means to catch them cheaply so that they can be exploited economically in large volume. . . .
. . . [The projected rate of growth in sales by domestic and U.S. owned overseas fisheries] . . . cannot be established or maintained, unless oceanic investigations are conducted on a world wide basis to find: (1) how locations and sizes of fish populations vary with the changing conditions of the sea; (2) the ocean conditions that bring about economically catchable fish concentrations; and (3) those aspects of behavior that can be exploited to reduce the costs of catching fish.[8]

Solely on the basis of those propositions and assuming that "products triple in value between producers and final consumers," the Academy report supposes that expenditures of fifty million dollars annually in oceanographic research will in ten years, through benefits to fisheries, add two billion dollars a year to the Gross National Product.

The *only* reasoned contention for a payoff of fisheries research is that a "rational development" of the American fishing industry can lead to the recapture of its rapidly declining share of the domestic market and, correspondingly, add to the GNP.[9]

The only conceivable way that such research could add to real per capita income would be by increasing the productivity of fisheries (hence, lowering real costs per unit of fish product) redounding to the benefit of the consumer as lower prices. Only by increasing the productivity (lowering the costs) of

American fisheries relative to foreign ones, could American fish production be expected to replace foreign imports. How then is it to be established that federally sponsored oceanographic research will increase productivity (or by how much it will increase productivity) for fisheries in general or for American fisheries in particular, or that it will give a cost advantage that will go to American fisheries relative to foreign fisheries.

Why should it be supposed that federally sponsored research in oceanography will redound to the special advantage of American fisheries? Scientific research in oceanography is not analogous to a factory, whose outputs are used up by the local consumers. Whatever is discovered through publicly sponsored oceanography concerning the ecology or biology of the organisms supporting marine fisheries, about how fish populations vary with the changing conditions of the sea, or concerning aspects of fish behavior will be as relevant to the problems and as usable by the fishing fleets of the Japanese, Russians, Germans, and British as by our own; just as, presumably, the published results of the oceanography of the Japanese, Russian, British, German, French are available to and possibly used by our own fishing industry. Whether the output of American oceanography will be more to the advantage of American fisheries or to foreign fisheries depends on the relative responsiveness of the American and foreign industries to research-based data—will depend, that is, on which is better organized to comprehend the relevance and to apply the results of this particular brand of esoteric information. It would appear that in recent decades the foreigner has been more progressive with respect to, and is more capable of capitalizing on, the benefits offered by such research. Inasmuch as this is the case, the result of more American oceanography will be to increase, rather than to diminish, the comparative advantage of foreign fishermen. The comparative advantage of American fisheries will *not* be augmented by more federally sponsored oceanography unless the industry is radically reorganized to make it more responsive to, and more capable of, innovation in exploiting the outputs of world research. About the present organization of the American industry, about its responsiveness to research outputs, about the ways and means to its reorganization and reform, the Academy report offers no clues.

But the most unfortunate aspect of the Academy study, and the one that renders it substantially useless, is its failure to establish any measurable

relationship, indeed, even to establish that any positive or viable relationship exists between oceanographic research and the productivity of fisheries.
The record of the research achievement of the already vast U.S. oceanography programs is surely available. But that record was not examined, nor was
any effort made to measure the concrete impacts and benefits of the research
expenditure already made. Surely, it is impertinent to predict what the
payoffs in oceanographic research will be in the future without glancing
back to see what they were in the past. Nor, alternatively, did the report
specify the research plan or the problems that the research would undertake
to solve, nor the information it would try to find nor try to establish the
probabilities the information could be found and that those problems could
be solved, nor the importance of solving the chosen problem, nor the value
that the sought for information might have if it were found. That would
have required going beyond the simple a priori science-input, progress-
output assumption to examine the subtle and equivocal relationships between research and information, between information and action. Instead,
there was offered a general discourse on the values of research, which could
have as well rationalized benefits of any imagined magnitude. Rather than
the cost/benefit ratio of 5.8/1, one could as well have proposed a ratio of
.058/1 or of 581/1. And why $50 million to be invested per annum? Would
the cost/benefit ratio differ if the investment were $5 million? or $500
million? The report offers no clue.

PART II. SCIENCE AND THE SYSTEM OF TECHNOLOGICAL ADVANCE

The System of Technological Advance

Suppose an established and continuing technology, i.e., an organized
capability for some purposeful action, learned by operatives, one from the
others or fixed in the social memory by reports, blueprints, machinery
designs, layouts, manuals, instructions, and so forth. What must go into the
process of replacing that established technology by another more productive one? Can we specify general and necessary components of change? That
process can be reduced to these essential elements:

> *Information,* broadly understood to include all that might be observed, com-
> prehended, realized, imagined, calculated, thought—and hence that is or

might be communicated. The process of change requires becoming informed concerning possibilities, potentialities, options for change. Information is what goes into "knowing about."

Innovation, experimentation, the "trying it out," the transition from information into organization and action. Innovation to determine the viability and values of information produces another order of information.

Transformation, reorganizing of operations. Innovation merges into transformation and hence might be considered as a first phase of transformation. But the purpose of and prerequisites for determining the ultimate viability of a change and those that determine the inauguration of the change itself, can be worlds apart and decades or centuries removed. Moreover, the transformation in one firm, industry, society becomes a source of information concerning the viability and value of change (hence, information, producing innovation) for other firms, industries, societies.

Communication. Whenever the entire process of technological change is not derived from the experience and contained within decisions of a single individual, there must be communication, through a variety of media, through teaching, publishing, discussing, demonstrating, selling, commanding, directing. Thus, schematically:

INFORMATION-	(COMMUNI-CATION)-	INNOVATION-	(COMMUNI-CATION)-	TRANSFOR-MATION
idea	speaking	experimenting	observing	reorienting
image	teaching	practicing	demonstrating	reorganizing
conception	publishing	measuring	directing	establishing
observation		evaluating	commanding	investing
invention				
data		testing	telling	
calculation				
possibilities				
need-demands				
availabilities				
options				
opportunities				
KNOWING	TELLING	TRYING	SELLING	DOING

What goes into any of these categories may be infinitely varied and complex; but, in every case, in searching, communicating, innovating, transforming there must be motivation coupled with the availability of resources to cover costs, coupled with the command (power) over those resources. Normally, the motivation and power to act is set in balance against a complex of inertial forces and countervailing motivations and powers.

Science in the System

This system of technological advance and each of its component elements deserves to be studied and needs to be understood in depth. Our purpose here, however, is only to locate science in the system.

Science, understood as a set of activities, is itself a system for producing information concerning selected subjects of inquiry. The information that science produces needs not have any relevance for or relation to technological advance. The information that sets in motion and sustains the processes of technological advance may be wholly outside the domain of science. Indeed the information that has been fed into that process as one of its essential ingredients has been, for the most part, generated by operating experience and observation, not by science and research. Nor need the information produced by science be "inventive," "creative," i.e., embodying a novel idea or conception or postulating and establishing a new set of relationships. Inventiveness and creativity are as rare in science as in other spheres of human endeavor. Scientists and everyone else, for the most part, learn through observation and experience in the frame of an accepted conceptual framework on the basis of given assumptions and pre-established relationships.

If it is but one of the systems for producing and communicating information, if it is but one of numerous sources of technologically relevant information, and if the information it produces need not be technologically relevant, then what gives to science its peculiar distinction?

"Classical" science is distinctive as an information system in the following respects: (1) it has a highly developed means of synthesizing, accumulating, and disseminating the information it produces; (2) credibility of the information it produces is established allegedly by experimental verification or at least by general exposure to the possibility of experimental refutation, with a motivation to test and refute built into the system.

At least a century ago science could properly be understood as an integral information-producing system having the aforementioned qualities. But, beginning with small deviations in the mid-nineteenth century, there has evolved two distinct systems of science, serving different clientele, having different structural characteristics, and operating on different bases, namely Academic Science and R and D (Research and Development).

The Qualities of Academic Science

Academic Science (which academicians prefer to call "pure," "basic," "fundamental") is in the classical mode.

(1) Like a church or a charity, Academic Science is valued as an activity of intrinsic worth, with the motivation of participants as to what they do and how they do it ultimately based on a dedication to internalized values at the institutional core.

(2) The information it produces is universally available, is actively disseminated through publication and through the teaching nexus, and is intrinsically valued as "truth."

(3) The subjects of its inquiry are chosen by the individual scientist according to his personal inclination, constrained by institutional values, the prevailing academic fashion, and certain collegial pressures.

The Qualities of R and D

R and D draws from the same conceptual and information base as does Academic Science, but it is not organized in the classical mode.

(1) R and D is not a self-contained institution valued for its intrinsic worth. It is an agent and instrument of the political authority or of business enterprise. As such, it is primarily valued for the contribution it makes to political purposes or to business profits.

(2) The information that R and D produces is not universally available. There is no built-in motivation to communicate and disseminate that information. In large part, the communication and dissemination of such information is closed off for reasons of commercial or political advantage, vis a vis competitors and rivals.

(3) There is, in general, no built-in motivation or institutionalized organization to select the information produced through R and D for its general values, nor to accumulate and synthesize such information, nor to feed it back into the teaching nexus.

(4) The problems of R and D are not selected according to the personal inclination of the scientist or by reference to some general criteria of worthwhileness. They are selected with reference to the purposes and problems

of the particular business firm or public agency to which the R and D operation is attached. They may be posed by an authority external to the R and D activity, or at least they must be approved by such an authority.

What has been said here is intended to compare the modal qualities of Academic Science and R and D. There are surely intermediary forms and variants. Nor should it be supposed that the characteristics ascribed to the two information-producing systems cannot be or will not be changed. On the contrary, they can be and are being changed.

The Role of Academic Science and R and D in Technological Advance

So far as the system of technological advance is concerned, R and D., in contrast to Academic Science, has the following great advantages.

(1) It focuses on technologically significant problems, and, on that account, the information it produces is likely to be more relevant to technological advance.

(2) R and D is organizationally connected to decision centers having power and motivation to organize the transformation of technology. Hence, back-and-forth communication will be facilitated (or necessitated) between those who specialize in producing information and those who need, can use, and are applying such information. On that account, R and D can be effectively geared into each phase of the system of technological progress.

On the other hand, inasmuch as the information produced through Academic Science is relevant to technological problem-solving and change, it is likely to be more generally available and effectively and universally disseminated than is that of R and D.

Oceanography in the United States has been organized generally as an Academic Science. Fisheries research approximates the model of R and D. Both, however, are normally under governmental auspices. Hence, the former is more directed than its university equivalent, and the latter need not be directly geared into operations.

Normal and Revolutionary Science

The modus operandi of Academic Science is not like that of the business firm nor of a political government. It is nearer to that of a church. Academic Science, indeed, is akin to religion inasmuch as its initiates find in it a set of goals, values, and a way of life that are intrinsically satisfying, worthy of dedication, needing no justification by reference to values or benefits ulterior to the science activity itself. On the other hand, as is usually the case with a church, science requires the support of outsiders who do not participate in its mysteries—who do not thrill to its adventure—simple believers who make their donation in the expectation of positive benefits accruing to themselves; for them, the mythology that sees in science an ever-flowing, overflowing horn of plenty. While R and D has made, and as a condition of its continued existence is obliged to make, contributions to technological advance, the bulk of Academic ("pure," "basic," "fundamental") Science has not and is not likely to make any such contribution. The claim of Academic Science to public support on the basis of its practical values must be made on another ground.

Both Academic Science and R and D share what has been called a "paradigm" of analytic method, conceptual outlook, modes of experiment, and prevailing theory. Nearly all science, nearly all the time, operates and generates information within the frame of that paradigm. Such is the normal path of Academic Science and R and D. Occasionally, science shatters its paradigm and creates another, as a new base for normal research. Such "revolutionary" science, whatever its source,[10] widens the scope and increases the effectiveness of R and D as well as of Academic research. It is in the shattering of the old paradigm and the construction of the new that the benefits of Academic Science can spill over into R and D and its production of information geared to technological advance, or conversely where the values of R and D spill over into Academic Science in answering questions of interest to academics.

There is a priori reason to suppose that Academic Science, dollar for dollar, man for man, is more likely to produce scientific revolutions than R and D, (only) inasmuch as the academic scientist is free in his choice of problems and paths of inquiry. For that freedom should enable the maverick and the creative rebel to pursue to the end his sense of the inadequacy of the authoritative outlook, persisting until he brings the structure down. In recent years, Academe has neither been free nor productive of scientific revolutions. Financed, project by project, through government and foundation grants, it too has served a paymaster. Its paymaster has not been the businessman nor the politician but academic authority, dominating the committees and ruling the foundations. Academic authority, it turns out, is invariably the enemy of scientific revolution.

PART III. SCIENCE AND FISHERIES

The Industry: Facts and Anomalies

The vast ocean space captures the imagination. Here is the frontier of technology, an immense resource that we have scarcely learned to use. For industry, the oceans have been an enormous sewer, a vast dumping hole. For the fisherman, they have been only a hunting ground.

Nevertheless, the increase in the world output of fish, from an annual yield of four million tons in 1910 to sixty million tons in 1967, has been very rapid and, save for the interruptions and destructions of World War II, at a quite constant rate.

Table 11
WORLD FISHERY YIELDS FROM 1900 TO 1967
(in million tons)

1900–10	4	1957	31
1924	10	1959	35
1932	10	1960	38
1934	14	1961	42
1936	17	1962	45
1938	21	1963	46
1947	18	1964	52
1948	19	1965	53
1953	25	1966	57
1955	28	1967	60

SOURCE: Meseck, *"Die ernahrungswirtschaftliche Bedeutung der Fischerei unter weltweiten Aspekten,"Berichte über Landwirtschaft,* Bd. XLIV. 1966. Heft 3, S. 518, and FAO, Fisheries Statistics, 1966, vol. 22

From 1948 to 1968, the annual yield of the world fisheries more than tripled, supplying a proportion of the human consumption of animal protein estimated at between ten and twenty percent.

The rate of production increase in the past decade . . . was far above the rates of increase in agriculture and forestry. In the world fishery yield of 1966 of about 57 mill. tons, sea fishery had a share of about 88% and freshwater fishery of about 12%.

According to calculations made by FAO, fishery yields contributed 10.5% of animal protein to the supply of the inhabitants of the earth. Some authors even quote this share at 12.5 to 15% and, after conversion to essential amino acids, at 20%. It should be emphasized that in the so-called "low calory countries" 21% of the available quantity of animal protein is formed by fish. In densely populated East Asia (1.7 billion inhabitants) even 50% of the daily ration of animal protein consists of fish. From this the eminent importance of the fish supply in this region can be seen.[11]

The changes in the fisheries yields by continents from 1938 to 1966 is shown in Figure 3; in Figure 4, by the largest producing countries; in Figure 5, by selected Asian LDCs.

It should be noted that by far the greatest increase has not been in the rich, advanced societies of Europe and North America, but in Asia and Latin America. In the country breakdown, the most spectacular increase has been in Peru, from virtually nothing in 1948 to approximately nine million tons in 1966, based on the development of industry from the *export* of fishmeal exploiting the tremendously rich fisheries created by the up-swelling water on its coast. Among the countries maintaining large seagoing fleets, it has been those capable of highly advanced technology, where the level of per capita income remains relatively low (Russia and Japan), where the increase in annual yields has been the greatest. On the other hand, in the United States, technologically still by far the most advanced country in the world, there has been an absolute decline in the annual catch since 1956. The same has been true in Germany, where the landings of deep-sea fisheries decreased by twenty-three percent from 1955 to 1966.[12]

It is commonly alleged that "the enormous increase in fishery yields during the past two decades is without doubt the result of scientific and technological progress."[13] Then how does one explain that the increase in fishery output is more rapid in the poor and backward countries, and that the fisheries output of the most technologically advanced countries in the world has declined absolutely and fallen far behind relatively, with great increases in output occurring elsewhere?

The anomaly can perhaps be explained as follows:

(1) The great increase in the output of the world fisheries is not the result of any significant technological advances nor of the input of scientific information, but rather of the extension of established activities and the increasing rationalization of those activities, sometimes following foreign models, among hitherto traditional societies. Thus, for example, the production of

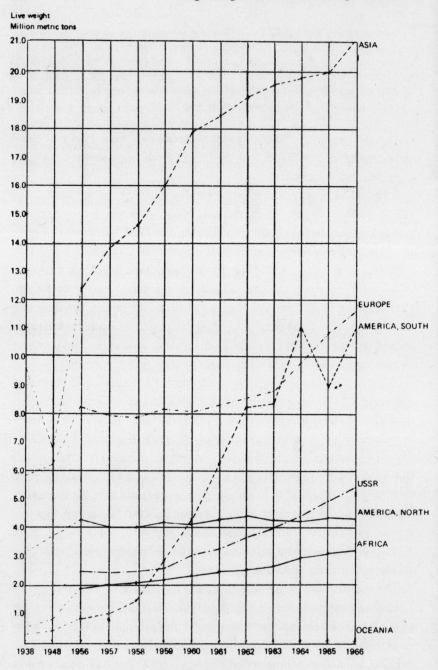

Fig. 3. World catch by continents.

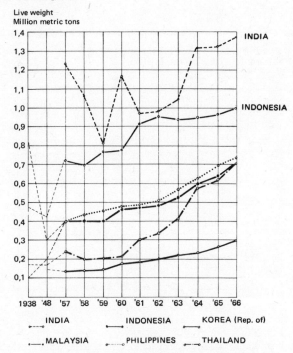

Fig. 4. Fishery yield of several Asiatic countries.

fishmeal in Peru or the use of trawl net fishing in Thailand, accounting for significant increases in world fisheries output, are in no sense a function of scientific research or technological novelty.

(2) The Germans suffered a destruction of ports, ships, and landing facilities during World War II, the loss of fishing areas as other nations pre-empted offshore rights, and a decline in the fish population (the reason for which is not understood) in the North Sea, e.g., the winter herring fishery of Norway declined from 1,146,000 tons in 1956 to 69,000 tons in 1961. This does not explain why the Germans failed to extend their activities elsewhere.

(3) In the American case, fisheries, like construction, fall into that intermediary zone of small, craft-based enterprise, outside the active intervention and technological and research support of the political authority and outside the potential for technological rationalization and the use of R and D characteristic of massive corporate enterprise. Industries of this intermediary zone, detached from government as well as from corporate R and D, drained of engineering and scientific talent, have been technologically

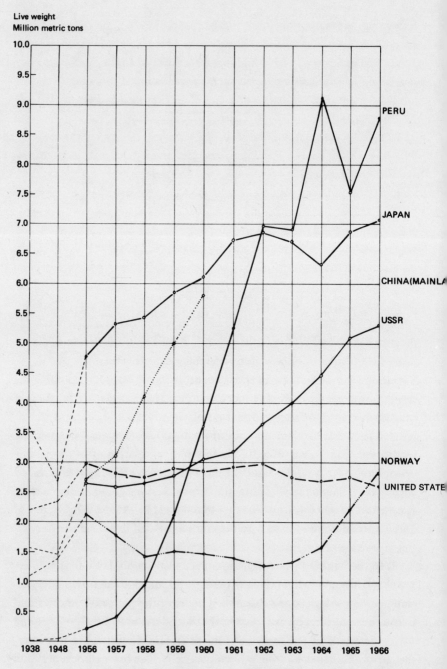

Fig. 5. Aquatic animals and plants: catch of the 6 largest producing countries.

the most laggard in the American economy. In the United States, moreover, the immigrant heritage of craftsmen has been depleted, with no effective system for development and renewal; craft-based industries like fisheries have, on that account also, fallen technologically far behind the rest.

But this still does not explain why the fishing industry was not transformed to a massive, mechanized enterprise as has evidently been the case in the Soviet Union and Japan. There is another explanation, applicable to both Germany and the United States, and generalizable to the state of advanced technology and high per capita income.

(4) Technologies may shift to technically less advanced and less productive economics and operate there at a comparative advantage for a number of reasons. One reason surely is that the technology in question is, so to speak, self-contained and can be detached from the matrix of complementary industries and social organizations with a minimal effect on operations. Of this, the deep-sea fishing operation is an excellent example. The vessel on the high seas is a self-contained operation, not a cog in a vastly larger sociotechnical complex. The American or German vessel lacks, therefore, the great advantages accruing to another sort of enterprise that is *embedded* in and is an integral part of the orderly and powerful sociotechnical complex of the U.S. or German economies, relative to the same enterprise encased in the sociotechnical matrices of some other, less well-organized, less advantaged national economy. Reflecting this, German and American operations, self-contained on the high seas, are bound to enjoy a lower productivity differential vis-à-vis the foreigner, than does their own "embedded" enterprise. Supposing that wages in each industry are equal to the average (national) productivity of substitutable labor in the economy, and that the wage/productivity ratio is a measure of the real wage of labor, then real labor costs will be relatively higher for seafaring, hence this (or any self-contained form of) operation will be comparatively disadvantaged by the high degree of technological advance for the economy as a whole. Accordingly, the gains of trade would be optimized by a greater share accruing to relatively less-advanced economies.

Oceanography and Fisheries R and D

During the early 1960s the massive international expenditures on ocean-ography was heralded with bright promises of benefits-to-come. Vast mineral resources were to be found and exploited. Mankind was to move from the stage of a sea hunter to a farmer and cultivator of the sea. A decade later, those promises remain promises, as far away as ever. And one can ask: what have been the practical benefits of oceanographic research? It is not clear that there have been many.

Indeed we may wonder what is the subject of oceanography? The title is as broad or inclusive and empty as an above-the-sea-ography or an in-the-air-ography.

What oceanography has attempted evidently are multidimensional surveys of oceanic areas, to determine the topography of the ocean floor, the thermal structure and the chemistry of the waters, the dynamic patterns of flow and change. Those surveys attempt to find out and to set down "what is there," and there is no end to what is there. By the ethic of science and by the ethic of modern man, it is a good thing to know "what is there." But useful? So far the information has been little used. Nor is this surprising. The universe under the sea, as the universe above the sea, is infinitely complex. It is, therefore, unlikely that any set of facts collected concerning it will be useful for X unless they have been sought for and collected with the interests of X, X's purposes, X's problem, X's needs and circumstances in mind. And the surveys of oceanography have not been oriented to serve the interests, to satisfy the needs, to solve the problems, or to bear upon the policies of government or enterprise. Will this systematic laying out of the facts, beyond a knowing more about, lead to a new and deeper understanding of oceanic phenomena? to a paradigmatic change? to a scientific revolution? Perhaps they will, but they have not as yet.

What then has R and D done, vectored on the specific needs of fisheries, and what can it do to promote technological progress?

We are still hunters of the fish of the sea, increasingly numerous and destructive hunters. Given that status, technological advance takes the form of more efficient hunting through: discovering where the fish can be found and developing the instruments and technique for catching them, and after that for preserving, transporting, and distributing them more easily and cheaply. Both the exploration for fishing grounds and the development of techniques and instruments for catching, transporting, and preserving the fish catch, require systematic inquiry, knowledge, and experiment. Where large and modern fishing fleets exist, it is they who are best equipped to explore, and have in fact most effectively explored the sea for new fishing areas. The development of techniques and mechanisms for the catch—e.g., the use of nylon and other synthetics in the fabrication of lines and nets, the use of sonar, and radar, and television to locate and identify schools of fish, the use of powerful light beams as lures (in purse seine and basnig fishing); and for the preservation, processing and transportation of fish, e.g., deep freezing and mechanical processing on factory ships at sea—have indeed required imagination, engineering skills, and a practical knowledge of fishing and of the fishing industry, but they hardly require the work of the research scientist. These instruments and techniques have, in fact, been developed out of the experiences of commercial enterprise, the fleet and the industry, with fisheries R and D sometimes playing a collaborative role. The distinctive contribution of fisheries R and D has been in the systematic organization of information concerning these developments and its orderly dissemination particularly through vocational education and training, as well as through programs of technical assistance to countries without modern fishing fleets and modern fisheries-oriented enterprise.

Beyond what goes into catching, preserving, transporting, distributing the fish, a rational exploitation of the fisheries resource requires information on the effect of the current catch on future yields, for two purposes: (1) to find out the short-run relationship between the size of the catch and the marginal costs of incremental fish yields as an important datum in the economics of fishing fleet operations and (2) to find out the long-run relationship between the magnitude of the fish harvest and the capacity of the stock to reproduce itself as a basis for the determination of safe limits of exploitation and for the design of systems of control. In the dual purpose study of the effect of the current catch on future yield, fisheries R and D, under public auspices, must play a central role. In respect to this task, it should be noted:

(1) The determination of safe fishing levels or of optimal catch is very difficult and bound to be controversial, inasmuch as reproductive capacity of fish species and the size of the stock is affected by numerous variables

other than the size of the annual catch and inasmuch as there are complex ecological relationships between species so that a change in the number or condition of one species acts upon the numbers and condition of other species.

(2) More than the determination of an optimal level or safe boundaries for the annual catch is required for a system of control. There must be the motivation, the agency, the power to control. For the open seas, beyond all national dominion, an agency of control is difficult to establish, and the power to control is difficult to mobilize, particularly since motivation is equivocal inasmuch as the individual fisherman or the fishing nation that reaps the gains from overfishing need not be the one who pays the cost, i.e., the general "externalization" of the diseconomies of overfishing. This is a central fact in every effort to rationalize the use of the resources of the sea.

Probably, as the size of the annual harvest is pushed higher and higher, and safe fishing limits are reached and surpassed, the sense of mutual loss will override the greed for individual and national gain, and control arrangements will be made. In this process, the R and D institute will hopefully play a salutary role in articulating the need, in acting as a political pressure-point for change, and in designing and negotiating control arrangements.

Grazing the Sea

It is often said that fishing techniques have not advanced much beyond mankind's hunting stage, and that what is needed is to farm the ocean. However, the problems of fisheries are more properly compared to cattle grazing on an open range than to farming. A balance between catches of different kinds of fishes must be maintained, or else the kinds that are least useful will take over from the most wanted varieties. We need to learn how to breed better varieties of fishes, like salmon, that fatten themselves in the distant seas and return to the rivers to spawn, and how to control the predators and pests that compete with us for the harvest of the sea. The addition of small quantities of vitally needed substances may increase the fertility of the ocean pasture, in the way that the Australians have been able to improve their sheep range by adding small quantities of cobalt.[14]

As a next step in rationalizing the use of its resources, the sea might be organized as a grazing area rather than as a hunting ground through discovering or developing and protecting breeding grounds and sea hatcheries, protecting valued species against predators, influencing or producing the upwelling movements of the currents, or otherwise fertilizing the surface of the seas.

A word concerning the "fertilization" of the waters. Fish need food; and, other things being equal, where the food is, the fish multiply. Food for fish populations is produced by vegetable organisms near the surface of the water. These, through photosynthesis, capture the energy of the light and in combination with the minerals of the sea form a nutritional substance. Over great areas of the oceans, particularly in tropical seas, there are relatively few fish because there is relatively very little of such vegetable matter growing on the ocean surface. The reason for this is a paradoxical one. Because of the intense sunlight, photosynthesis can take place in these areas very rapidly and at a considerable depth producing an abundant growth of vegetable matter. This would seem to favor the increase in fish populations, except that the rate of photosynthesis is so rapid that the consequent consumption of mineral matter from the surface areas of the seas overbalances the normal inputs of minerals into the upper surface. Hence, these surface waters, thinned of their mineral content, can no longer support any but a meager growth of vegetable matter. The rapid depletion of the mineral content of tropical soils (described earlier in the chapter on science-based AID in France) presents an almost exactly equivalent problem.

The exception to this rule are those places in the tropical oceans where peculiarly strong currents rise from the mineral-rich sea bottom to renew the mineral content of upper levels. It is those areas of ocean upwelling precisely where one finds a remarkable growth of sea vegetation and large fish populations.

Large populations of fish and invertebrates are very often associated with regions of upwelling. The enormous population of anchovies off Peru is substantially confined to the upwelled waters there; off Dahomey, where the water upwells behind a cape, 5,000 canoes may take 10,000 tons of sardinella during a two-month season. The pink shrimp of the Gulf of Panama come in with cool, upwelling water; the king mackerel *(Scomberomorus)* schools at the surface in the Gulf of Aden when the surface waters are cooled by upwelling. Along the Sarashtra coast of India, cool subsurface water creeps up over the continental shelf as the monsoon and the surface current change. It is then that the great Indian salmon or Dara *(Polydactylus)* swarms into the stake nets.[15]

Some have speculated on the idea of building atomic piles at the ocean's bottom so that the heated currents rising would carry mineral-rich waters from the sea bottom to the sea surface.

There may be other simpler, less dangerous, more practicable means of dealing with the problem of mineral-thin surfaces of tropical waters. Just as is done on the land, the water surfaces also could be fertilized. Imperial Chemicals, we are told, tried this with spectacular results in the northern fastness of a Scottish loch; but in the tropics where environmental circumstances would favor such an approach, no such experiments for ocean waters appear to be in process or in the offing. Of course, a practical payoff does not simply depend on the proven feasibility of multiplying fish populations. Lacking arrangements for an international sharing of costs and benefits, it depends also on finding enclosed (or in enclosing) ocean spaces where the value of an incremental fish supply can be captured by those whose investment has produced it. The costs, and hence the economic feasibility of such a venture, will also depend on the proximity and availability of whatever mineral-rich materials could be used as fertilizer and on the development of techniques for fertilizing the waters.

Should the sea be organized as a grazing area, fisheries R and D under public auspices would certainly play a big role. So far the contribution of such R and D has been exploratory and suggestive, e.g., studies in Britain on the effects of protected environment for hatching ocean fish stocks. It is unreasonable to expect that fisheries R and D will ever gear itself to the development of the oceans as a grazing area or will produce information of more than marginal relevance to that task *unless* there is an agency of decision and action demanding such information and able to use, and using, it. There is no such agency. The effort to create one would require motivation and power, which now is lacking, and would confront all the problems of the multinational organization of activities where benefits may easily slip from the grasp of those who pay the costs.

Farming the Waters

It would be wrong to say that man has not yet reached the stage of the fish farmer. On the contrary, the culture of marine animals (mariculture) has been carried on for many centuries. About twelve percent of the annual fish yield is from inland waters and, of this, a considerable proportion is the product of mariculture.

In Europe, there are trout pools in Denmark, German eel pens in the North Sea, and especially the cultivation of mussel (a delicious morsel that is allegedly the most efficient protein producer of all) in Spain, France, and Holland (283,000 tons in 1968). Mariculture has been especially important among the ancient societies of Asia: Japan with its shrimp culture, China which produces perhaps 2,000,000 tons of fish annually in ponds and paddy fields.[16] In 1967 Indonesia farmed 138,000 tons of fish; Taiwan, 56,000 tons; the Philippines, 65,000 tons.

Mariculture has evolved as an ancient craft—complex nevertheless, requiring patience, skill, knowledge—in Europe as well as Asia, where (for example) mussel culture was first developed by a shipwrecked nobleman in the thirteenth century. Methods of cultivation are various. The mussel is cultivated on strings suspended from rafts (floating gardens) in Spain; on posts anchored parallel to the coasts in France; in sea bed gardens in Holland, England, and Germany, always washed and fed by the movement of the tide. The marine animal may be suspended in flowing water, carp in cages in the polluted rivers of Indonesia, oysters on ropes in the ocean waters of Japan consuming the through-put of plankton. The marine animal may be cultivated in brackish or freshwater pools or in paddy fields, fed directly, or feeding on vegetation growing at the bottom or on the water surface, variously protected from predators. Practice and results in the more technologically sophisticated Asian economies (Japan and Taiwan) establish the great increase in yield that is possible through improved technology.

The famous milkfish farming is popularly practiced in the Philippines (140,000 ha.) and Indonesia (125,000 ha.) and has a history of over 600 years. Unfortunately, throughout the centuries little improvement has been accomplished, and the techniques used at present remain mostly traditional and empirical. The average annual production is only about 400–500 kg./per ha.

Improved techniques developed and practiced in Taiwan have succeeded in producing as much as 2,500 kg./ha./ann. These improved techniques introduced to the Philippines recently have already shown spectacular results. In the localities where such new techniques are now practiced the average annual production has already increased by about 100 percent, from the old 500 kg./ha. to about 1,000 kg./ha., and is expected to continue increasing to 1,500 kg. or even 2,000 kg./ha. in the near future when pond operators become more experienced in practicing the new techniques. When sufficient numbers of extension workers are available, such new techniques are expected to be popularized and practiced by all the pond operators in the country.[17]

. . . the prawn ponds of Singapore, which are known to have the highest production in this region, are producing only about 300–400 kg./ha./ann., while the prawn ponds in Japan, operated with improved modern techniques, are able to produce up to 3,000 kg./ha./ann.[18]

. . . there are many promising possibilities for development of a genuine marine agriculture, or "mariculture," in shallow near-shore and semi-enclosed waters. In recent years, Japan has been the world leader in this field, culturing seaweeds, particularly "laver," oysters and other molluscs, yellow-tail, shrimps, etc. In oyster culture, the Japanese have succeeded in increasing the productivity of a hectare of sea area from 700 kilogrammes to as much as 35,000 kilogrammes, a fifty-fold increase, by attaching the seed oysters to long ropes hanging from rafts. In this way, bottom-living predators can be avoided, and the entire volume of the sea in the area of cultivation can be utilized.

In agriculture on land, plant nutrients are added in the form of chemical fertilizers. In Japanese aquaculture, marine plankton are the "fertilizers" that nourish the crop of shellfish. The areas most useful for aquaculture are not those that can be enclosed or fenced in, but rather those in which there is a continual through-put of plankton carried by currents past the location where the crop is being grown. Substantial production of aquatic animals through mariculture is still in a primitive stage.[19]

To upgrade technology, what is required is to a considerable extent simply the rationalization of traditional practice.

The improved practice consists of the following main points:
(1) improvement of construction of ponds and their bunds, water gates and water supply and drainage system to insure good water supply and efficient control of water level and to minimize flooding or leakage;
(2) promote production of natural food supply—the complex algal pasture at the bottom of the ponds—by application of fertilizers and management and control of water qualities to insure a sufficient and continuous supply of rich food;
(3) maintenance of optimum density of population (in terms of total weight of fish per unit of pond) to insure optimum utilization of available natural

food supply by stocking manipulation with fish of different size groups and by repeated harvesting (selective harvest of the marketable size group) followed by repeated stocking with the youngest size group;
(4) strict prevention and control of predators and competitors;
(5) improvement of fry collection and nursery operation to insure sufficient supply of stocking material;
(6) improvement of harvesting gear and technique, and care, handling and transportation of harvest.[20]

In the instance of shrimp, however, the Japanese "through many years of patient research and experimentation" have developed techniques for the scale production of seed prawns, replacing the traditional methods of collecting them from their natural habitat. To transfer this very important development would require that it be adapted to species flourishing under conditions prevailing elsewhere.

Success in increasing shrimp farm production depends mostly on the availability of supply of shrimp seeds. Natural occurrence of young shrimps and shrimp larvae fluctuates greatly and is often unpredictable and to collect them in large numbers from their natural habitat is difficult and expensive. An adequate supply of the stocking material can be assured only by producing them under controlled conditions.

Through many years of patient research and experimentation, efficient techniques for producing seed prawns of *Penaeus Japonicus,* in large commercial scale economically, has already been developed in Japan. . . . However, such techniques developed in temperate regions based on temperate species need to be experimented with and tried under Southeast Asian conditions and with local species, modified and improved accordingly . . .[21]

Not only is there this great potential increase of productivity within grasp, there is also is the great potential for extending mariculture even as it is presently practiced. But if the potential increase in production is enormous, enormous also—given that cultivation is by millions of peasant farmers spread over wide areas and probably outside the modes of modern thought—are the problems of communication, dissemination, and the organization of technological change, as well as in the collection, processing, and distribution of marketable surpluses.

In inland fishery the present world yield of freshwater fishes could be duplicated solely by systematical utilization of the rice fields (especially in Asia) for fish culturing. But also in all other subtropical and tropical regions fish culturing, on account of the rapid growth of some fish species, offers manifold possibilities. On the whole, however, the economic mobilization of inland fishery will, on account of the production in single places far distant from each other, be much more difficult than that of sea fishery, which offers more favorable starting points for common organization of the fishing fleets, concentration of the transloading operations, and distribution of the catches.[22]

It is understandable that development planners and the agencies of public action in the poor countries of Asia have preferred to concern themselves with the development of modern commercial fishing fleets, which are so much more amenable to the planner's directive and to the banker's calculus. So also, R and D orients itself to the demands and problems of the action agency.

Training and educational programs turn to where the research is and where high-grade employment opportunities now are—neglecting mariculture, though that is the sphere of fisheries development where the greatest potential for technological advance must surely lie. S. W. Ling remarks,

> . . . a large number of persons have received general training in fisheries but few, if any, have adequate specialized training in brackish water fish and shrimp farming. There are fisheries schools in the various countries of the SE Asia region, but none offers extensive courses in this field.[23]

Mariculture has been virtually outside the scope of technology and the concern of science in the United States. There is now a problem of increasing urgency in the United States and in other affluent nations which may, and should, change all that, and draw the focus of research directly on mariculture and related phenomena. The problem is that of pollution.

Pollution of the Seas and Waters

A word about pollution in general before considering the pollution of our rivers and the sea.

There is a pollution fad. That is not to say there is no problem of pollution. There is also a cure-for-cancer fad. That is not to say that there is no urgent need to seek a cure for cancer. It does mean this: pollution has become a popular cause, a "crusade."

We carry in our minds quite different images of the crusades. For some it is of the handsome knight, eyes tenderly uplifted, faint halo round his head, kissing a cross. It can also be an image of a simple-minded mass of enthusiasts, excited by symbols and slogans, led by professional preachers and fanatics, none with an idea of cost or consequence, off to slaughter and be slaughtered until at last the enthusiasm has worn itself out. Whatever the nature of the historical crusades, it is that latter image that corresponds to those recurrent crusades in American life—masses of enthusiasts looking for ultimate solutions and easy answers, rallying around symbols and slogans, led by professional preachers and fanatics, who have no notion of costs or consequence, off against vague and distant enemies or to find a holy grail, until the enthusiasm is exhausted and the crusade fades away.

The concern with pollution may be new. The phenomenon is not. Nor are the great cities of the world, filthy as they are, as polluted as they once were. The Black Death, the bubonic plague, typhus, that wiped out great populations, were all products of pollution in the past.

Yet perhaps pollution in the modern world—the pollution of high technology and of urban density, of synthetics and atomic energy, and of jet- and auto-borne travel—creates a problem of another order and of a fundamentally different sort than that which we have known before.

Kenneth Boulding observes:

> We are now in the middle of a long process of transition in the nature of the image which man has of himself and his environment. Primitive men, and to a large extent also men of the early civilizations, imagined themselves to be living on a virtually illimitable plane. There was almost always somewhere beyond the known limits of human habitation, and over a very large part of the time that man has been on earth, there has been something like a frontier. That is, there was always some place else to go when things got too difficult, either by reason of the deterioration of the natural environment or a deterioration of the social structure in places where people happened to live. The image of the frontier is probably one of the oldest images of mankind, and it is not surprising that we find it hard to get rid of.[24]
>
> . . . Earth has become a "space ship" and a very small, crowded space ship at that, destination unknown. Up to now the human race has behaved and acted as if it lived on an illimitable plane. Now the earth has become a sphere, and we have to think of society as a sociosphere, that is, a sphere of all human interaction; and in this respect a sphere is very different from a plane. The great plains are gone for good; that was an episode, and a very brief episode, in human history, and we will probably never be able to go through it again.
>
> In the space ship there are no mines, no ores, no fossil fuels, no pollutable reservoirs, and no sewers. Everything has to be recycled, the water to go through the kidneys and the algae to the kidneys and the algae, and so on

indefinitely. Everything has to go from man to his environment, from the environment back to man. This is a very different kind of economics from that with which we are familiar. It is an economy in which the overriding consideration is parsimony in consumption, not the expansion of consumption. From now on the space ship is going to begin to close in on us. From the point of view of pollution, this may be much closer than we think.

We are already producing irreversible changes in the atmosphere which are causing alarm among the meteorologists, and it is clear also that we know very little about what we are really doing, that we do not understand the earth at all well, and that the earth sciences, even the physical sciences, are shockingly backward. It may be, of course, that for the present generation or two this is simply a problem in economics. We have to manipulate the rewards of the system so that pollution is not rewarded. It is a problem, however, which may easily go beyond economics, simply because man has not yet learned to develop a technology which is really stable in the sense that he could live comfortably in the midst of a self-perpetuating cycle. Our present technology is suicidal.

We will certainly run out of ores and fossil fuels in historic time, and we may run out of pollutable reservoirs before we run out of mines. The growth of affluence, that is, may be limited sharply by the growth of effluence. There are subtle questions also about the transmission of culture and values which are worrying and about which we know very little. All societies produce effluents of people as well as of sewage, the criminals, the mentally sick, the self-perpetuating poverty subcultures, and so on. The more complex a society, the more prone it may become to human effluence; and we may learn that values have to be recycled just the way nitrogen does.[25]

We earth creatures live in and as a part of an extraordinary system of geophysical processes and photosynthetic interactions that has permitted, so far as is known uniquely in the universe, the evolution of man, indeed, of what we call "life" in any form. Call this amazing system maintaining an environment propitious to life in this tiny spot in a vast dead universe, a "life system."

Our life system is not fixed and irrevocable. It has not always been. It need not always be. It came. It can go. It evolved. It can devolve. There were very long periods of time when conditions on earth did not permit the existence of human, or indeed, of any form of life.

That life system that perhaps produces the only environment in all the universe compatible with human existence is not man's invention nor a product of his technology. It has indeed been outside the scope of man's comprehension. Throughout his whole period of existence, he has taken the life system for granted, as given, and has acted freely within and thoughtlessly upon that system. Within the system that permitted him life, he polluted, poisoned, and destroyed, annihilating other species and even threatening his own; but the system itself, the system producing a physical environment compatible with human life, seemed indestructible, or at least beyond the powers of human destruction. It is in this respect that the modern phenomenon of pollution differs from an older variety. So massive and powerful is our technology, so densely interactive is our population, so rapidly do we consume—that is to say, disarrange what the life system arranges, that we are now able to degrade and destroy the life system itself.

Pollution can be understood as a malevolent, in some sense dangerous disarrangement, of the beneficent inputs and even of the functional determinants of the life system.

Pollution is disarrangement. It is a function of our scientific, knowledge-based power to control, manipulate, utilize—hence to disarrange—what the life system arranges. Formerly, our capacity to disarrange was more than compensated by the vast equilibrating power of the life system to rearrange, to bring back balance. What, now, if this is no longer so?

Shall we slow the pace of human activity, disarranging (that is to say, utilizing) less? I think not. We are too far gone for that. We cannot cease to utilize or even to reduce the dizzy rate at which the utilization of resources—hence destruction, displacement, disarrangement, pollution—increases. What then?

In answering this crucial question, we must avoid the economists' trap, namely in conceiving of the social problem as being essentially one of rational choice between given alternatives. Social alternatives are not given. They must be created. Beyond choice, more fundamental and more distinctively human than choice, is the task of developing the alternatives between which we choose. The essential social process is not of choosing but of learning, and the most important choice is in choosing what to learn. There is the crux. The whole thrust of our cognition formation has been based on the notion that the life system will rearrange what we have disarranged, that the thrust of technology ended and could be allowed to end at the point of consumption, that anything extraneous to the act of consumption could be ignored, that the air and the water were natural sewers that carried out of sight and mind whatever was dumped into them. And if this is no longer so? Then we must learn to rearrange what we disarrange, to re-equilibrate, to reinforce the life system in its capacity to produce and maintain a benign environment on this earth.

Consider the pollution of the sea and the inland waters. It has been suggested[25] that there are five types of water pollution: (1) toxic materials, (2) inert materials, (3) putrescible materials, (4) heated effluents, and (5) radiation waste.

Consider these.

Toxic Materials. Obviously, poisons are poisonous. They kill insects and rodents, also birds and fish, not only in lakes and rivers but now also in the most distant ocean areas; and, in the flesh of fish, shrimp, lobster, oyster, the poison comes back to us. Substances such as DDT are not spontaneously and easily broken down into nonpoisonous forms. We can treat the water course to neutralize the effects of these compounds at a cost, and we can *learn* to reduce the cost of so doing. We can *learn* to produce and use substitute substances or other means of achieving the same end. These are purposes for R and D. And we will need a science-based system of surveillance to find the poisons in the rivers and the sea, and to trace them to their source.

Inert Material. If heaps of garbage, of old cars and plastic containers, of slag and abandoned autos are piled on a grass lawn, the grass will suffocate or it will at least be kept outside the reach of grazing animals. The same holds for a river, lake, or ocean bottom. Vegetation can be choked. Fish, unable to graze, can starve.

We can hardly run out of holes in which to dump refuse, since we dig holes in order to get the raw materials to produce what is junked. The question is only of the costs of carrying the refuse to where it is properly dumpable. And beyond a rational dumping and disposition of inert waste, is *learning* to rearrange, recover, reuse it at a lower cost and to our advantage—which is another task for R and D.

Putrescible Materials. The pressing problem in the United States is not with the disposition of poisons or junk but rather with a mass of substances that are closest to and merge most easily into the life system.

Organic materials and inorganic elements are dumped into rivers as sewage and factory waste. Normally, the life system absorbs these, arranging them as benign components of the environment. Through bacterial action, the organic material is broken down into its elements. These, along with inorganic minerals, "fertilize" the waters. Through the photosynthesis of water-growing vegetation, they provide nutriment for the fish population.

But water, like soil, can be malfertilized or overfertilized. The input of organic material increases bacterial activity. Increased bacterial activity and the intensified growth of plant life reduces the oxygen content of the water. Through a surfeit of fertilizing material, the oxygen can be reduced to the point where the environment ceases to be a viable one for fish species. The established life system is then thrown out of kilter. As the process continues, the oxygen can become virtually exhausted so that vegetative growth and bacterial activity also cease. The area becomes a dead place and a place of death. So allegedly some 2,600 square miles of Lake Erie, completely drained of oxygen, has become.

What is to be done?

Water treatment plants accelerate bacterial breakdown by aerating the water. We can *learn* to aerate, infusing oxygen into our waters, more effectively, more cheaply—renovating the waters.

We can *learn* to use degradable materials—sewage, detergents, factory wastes—selectively, deliberately, in order to fertilize ponds, lakes, bays, sounds, indeed any natural or artifically created body of water in appropriately developed systems for algae raising and fish breeding.

Ecological systems could be researched and devised, *learned,* that utilize this great efflux of life-producing (as well as death-inducing) materials. It is just as possible to cultivate water-growing as land-growing plants, and to feed fish species upon them, or to develop more complex ecological chains with small fish eating the water-grown vegetation and marine creatures, and larger fish living upon the small (as the salmon was brought into Lake Michigan to feed upon the alewives). Given the heavy inputs of water-fertilizing materials, producing a growth of underwater and surface vegetation so dense that it destroys (or threatens to destroy) the established ecological balance, if that balance is not to be preserved, then surely other ecological systems can be devised and installed, also beneficent, adapted to the new condition. Increasingly, one hears of early still very shallow research efforts in this direction, e.g., testing fish species, imported from China, that flourish in the thick undergrowths of algae.

Compared to Japan, China, Taiwan, the Philippines, Indonesia, we are a backward country in the technologies and skills of mariculture. It is the destructive force of pollution, befouling our waters, that now drives us to *learn,* to bring to bear our research prowess and technological ingenuity in learning to capitalize on a curse and to utilize an affliction.

Beyond fish breeding and grazing, it is indeed time we moved to the stage of mass aquaculture, truly selecting and developing for their qualities plants that grow in the water as on the earth. Sea weeds and water-growing plants are rich in nutrients. They have qualities and values yet to be probed. Like all organic matter they can be processed into a very great variety of usable products.

Why not floating factories to dredge, harvest, and chemically process the thousands of square miles of undergrowth that chokes Lake Erie? Because it does not pay. It does not pay because we have not developed a technology to make it pay. And we have not developed such a technology because there has not been any effort whatever to develop one.

The meaningful choice is not between existing options, but between directions of inquiry intended to produce options that will open up new avenues for reasonable choice. We must—time will allow us no choice— seek out and create those options that will permit us to enjoy and exploit our waters within the conditions of our dynamic powers to disarrange and to rearrange the life system.

Heated Effluents. A factory draws water from a stream, uses it to cool its heated machines or to produce steam as an energy conveyance, and then pours the hot water back into the stream. This threatens and can destroy the established ecological balance. Fish and organisms may not be ac- climated to, or able to bear, the temperature changes. The heat reduces the water's oxygen content with all the maleffects mentioned earlier.

Factories can be and perhaps should be required to store and allow their waters to cool before being flushed into the stream. It is also possible to turn the heated water to good advantage. It is possible (as ongoing German experiments have shown), by controlled re-entry of the heated water, or by staging its re-entry in a series of pools kept at different temperatures to develop maricultures of a range and on a scale that would not otherwise be possible.

Radiation Waste. Radiation waste means wasted energy. Of all resources, raw energy is the most ubiquitous. Why then must we waste this energy?

Why must it be wasted? Again the dialectic. It is wasted because the materials are hard to handle; uses for them have not been devised; it does not pay. It does not pay because the technology has not been created to make it pay. The technology has not been created because the required effort has not been made to create one.

Fisheries R and D and the LDCs

Fisheries research is one of the many small overlapping worlds within the large universe of scientific activity. It employs complex and specialized experimental facilities. It has only a small number of experienced and creative workers and evidently, only a very few centers of significant R and D performance. These centers have evolved as the consequence of a subtle evolution of relationships with correlated strata of political and industrial activity and choice. Communication between these research centers is not arbitrarily restricted nor are information-outputs systematically seques-tered as might be the case if military, or proprietary control of private ownership were superimposed—but communication and the flow of infor-mation is not entirely free and open either. No doubt the language barrier and divergence in research orientation tends to close off the inflow of research outputs from Russia, China, and from the most important of fishing nations, Japan, to U.S. or West European R and D.

Supposing one had the power to plan a rational organization of the world's science resources as a means of promoting the development of fisheries in low-productivity societies. The relevant resource—i.e., ex-perienced manpower and specialized facilities—is sharply limited and is already highly involved. At any moment of time that resource can only be marginally diverted to new problems and responsibilities. Because the com-plex of capabilities to be tapped are built into organizations and these organizations are physically located in various countries and are financed by different national governments, and since the problems on which they focus are normally regional (or "situational"), the reorientation of their capabilities in support of fisheries development among the LDCs would require coordinated planning and decision taking. The objective would be to draw these science-based organizations qua organizations into develop-ment programming, to give them the capacity to propose policy and to initiate action, with a continuing stake and motivation qua organizations in the successful culmination of the development program to which they are committed.

To organize, coordinate, and reorient a science capability spread through a number of countries certainly requires the support of national governments. But the coordination and control of such a program is not a task for politicians, nor diplomats, nor for the "house scientists" of state departments and foreign offices. It is a task for the R and D leadership of organizations in the relevant fields of capability, operating within political and financial constraints but having effective power to initiate policy and to organize its implementation. As any such program of development moves toward practical realization, it must involve additional science capabilities and other sorts of competency as well. This suggests the need for a multinational task force with continuing responsibility but with an internal composition that changes as the phases of the task are completed and the development program proceeds.

If the objective were to mobilize a research capability for developing fisheries in low-productivity areas, there would be no inherent difficulty in bringing together the leadership of the significant research institutes at least in Western Europe and the United States. They know each other well, and no doubt would welcome the opportunity to work together directly. Formed into a group, they might be asked to propose a plan to develop the capacity to exploit the fish resource of selected low-productivity regions, the plan to be submitted to their respective governments, or to an association of governments from whom financial support was expected. Assuming the governments agreed to the plan and pledged financial support, the affiliated fisheries-research institutes might then form themselves into a task force for the phases of its implementation, for example:

(1) Given the assurance of financial support for a long-range development plan, participating institutes could not only coordinate the use of their existing experimental facilities and manpower, they could expand their facilities and recruit and train new manpower for the projected developmental task, and for the future support of the yet-to-be established fishing industries. For the latter purpose, they would recruit and train "counterpart" scientists, research specialists, and technicians from the regions where development of a new fishing industry was intended. These would participate at each stage in the development program.

(2) A survey of fish resources would locate fishing grounds, determine the size of the stock, and suggest the safe limits of exploitation for relevant ocean areas. Available manpower and laboratory ships of the participating institutes would be shared and their use coordinated.

(3) The results of the coastal survey would be analyzed to determine whether the establishment of a fishing industry would be feasible and economically justified, and, if so, the form it should take. For example, a fishing industry might be developed at three different levels: (a) to supply coastal villages and raise the level of subsistence of households fishing for their own consumption, (b) to supply inland populations and urban centers adjoining the coastal region, and (c) for commercial export sales to world markets.

(4) The development of a fishing industry at any or at all of these three levels would require that manpower be trained to requisite skills, that a body of relevant information be acquired and relevant techniques be developed, that an appropriate infrastructure be developed, and that operating facilities be set in operation. For example, in order to raise the subsistence level of villages that are supplied by household fishing, it might be necessary to: develop appropriate small boats propelled by outboard motors or even in simpler ways and the requisite means for their production and maintenance; develop hand nets, traps and gear that can be fabricated by village craftsmen; devise navigational guides; develop techniques and simple equipment for the drying, curing and storage of fish in the village; chart offshore fishing areas; teach village manpower to use the techniques and to fabricate equipment thus developed. To create an industry able to feed inland populations, larger vessels must be provided. Professional maritime skills must be inculcated. Extensive information must be produced pertaining to the location and movement of fish populations and the behavior of the various exploitable species. There must be harbors and docking facilities, facilities for storage and processing, for the curing, cooling, freezing, or canning of edible fish and for the production of fishmeal and oil from the inedible portion. There must be roads and facilities for the inland transportation and storage of fish supplies, and facilities and an organization for the efficient distribution of fish to interior populations and of fishmeal to agricultural and industrial users, and even the development of new industries and processes that use newly available fish-based inputs. If commercial fishing for

export to the world market is to be the objective, then other facilities, skills and techniques, information concerning other fish populations, and perhaps the cooperation of international fishing and fish-merchandising interests would be required.

(5) A committee of associated fisheries research institutes could not be expected to plan the whole, or to implement every aspect, of such a program. Others would have to participate. Yet such a committee could provide the leadership in the continuous process of development planning and programming, since, with all their limitations, this group of scientists are uniquely equipped to muster dedicated and disinterested service, and knowledge of practical organization, techniques, and scientific information relevant to this particular developmental task.

(6) As a part of its planning and the implementation of its plan for the development of fishing industries in one region after another, the associated fisheries-research institutes would establish indigenous schools for fishermen in the developing regions. They would train teachers for and would provide for the supervision of teaching in these schools, and they would coordinate research operations, e.g., ocean surveys and fish searches, with their program for fisherman training.

(7) Also as part of its task, the associated institutes would plan and set in motion an R and D program designed to provide the requisite information and to develop techniques progressively to raise productivity in the newly established fishing industries. This R and D program would be carried out partly in their own laboratories and partly in research establishments set up in developing countries, initially as branches of the associated institutes and later as independent but still affiliated institutes. In order to train scientists and technicians for work in these new research agencies, the associated institutes, in the light of their own diverse capabilities, would set up an educational and apprenticeship program where recruits would be trained sequentially in several organizations so that they might learn through working in proximity to different sets of research skill and different research foci. Thus, also there might be created personal relationships between these newcomers and scientists in established institutions as a basis for future cooperation and ease of communication when the trainees later return to research activities in their own developing regions.

(8) Initially, the affiliated fisheries-research institutes would include the

important centers in the advanced industrial countries, and later the newly established centers in the developing countries. Existing centers would be only a nucleus for a much larger activity in both the industrially advanced and in low-productivity societies, as the organization becomes the spearhead of a significant category of research, education, and development planning and programming throughout the world.

A tentative blueprint for mobilizing the science resource as a means of developing fisheries among low-productivity societies has been suggested. This blueprint may be reasonable and workable, and hence may have an intrinsic value. But our purpose in proposing it was to illustrate an approach to development planning—one which would begin with related organizations and institutions of science (always supranational in their viable relationships and in the flow, impact, and significance of their outputs) and would build a development program on that base with the intention of enlisting the initiatives and orienting the capabilities of the science resource.

In fact, AID has been responsible for at least two important fisheries developments. There has been the creation of fisheries on the west coast of Africa under French aegis, in Senegal and the Ivory Coast:

> Senegal has considerable traditional artisanal fisheries. With the help of France, the various sections of this branch have been considerably promoted over the last years. Among other things, France contributed a great deal to the improvement of fishing gears and fish-processing operations. Furthermore, she intervened in the sector of marketing, by providing both instructors and capital resources. These measures taken by France and other groups, especially FAO, made it possible to increase the landings of artisanal fisheries from 26,000 tons in 1948 to over 100,000 tons in 1967. Thus, the Republic of Senegal is not only in a position to cover the national demand for fish but also to export considerable quantities without curtailing the national supplies.[27]
>
> . . . France has afforded a great deal of assistance [to industrial fishing in Senegal] but in a more indirect manner, viz., industrial fishery was initiated by the French with their own fishing craft on a private basis and on own account, which is partly still true nowadays. Senegal's advantage in this was that native fishermen, providing part of the crew of the French cutters, were in the course of time familiarized with industrial fishing practices and enabled to operate their own cutters later on, thus saving the cost of training for Senegal. Furthermore, France has acquired great merits by sending fishery civil servants to the Senegalese fishery administration. . . .
>
> France afforded the same type of assistance to the Republic of the Ivory

Coast, thus paving the way for a rapid development of the fisheries of this country. . . . The fishing fleet of the Ivory Coast increased from 2 units in 1950, to 80 units at present. This was not only a numerical increase but also a qualitative improvement . . . catches range in the order of 50,000 tons of fish per year. Apart from the measures taken by the French government, this success is to a great extent due to the private initiative of French fishermen. The French fishermen operating with their own fishing crafts even nowadays in the Ivory Coast do certainly not pursue their occupation in that country only in order to help the Ivory Coast but also to derive the largest possible profits from it. Their work is nonetheless of great value for the country.[28]

There was, secondly, the German introduction of trawler fishing in Thailand. This has had a paradoxical result. Before the introduction of the technique, the Thai catch was half that of the Germans; subsequent to its introduction, it rose to double that of the Germans. The fact is indicative of a larger event. The world center for fisheries is in Asia, among the modernized and traditional cultures of that continent, nations of fishermen and of fish eaters. And it is in Asia that the systematic development of fisheries technology through R and D and through formal training is also likely to center. While the quality of the research among the LDCs has been criticized,[29] frequently and no doubt justly, the number of scientists and trained technologists engaged in those fisheries R and D and the number of training programs, schools, colleges, university departments oriented toward the fisheries activity is already formidable.[30] And it has the telling advantage of being linked to a growing industry that is of critical importance to the economy and of central concern for government policy. Mutual assistance programs in which developing countries teach each other and learn together have already begun.[31] And, no doubt, all the world will soon benefit from that learning, as it deepens and grows.

NOTES AND REFERENCES

1. National Academy of Sciences, *Economic Benefits from Oceanographic Research,* publication 1228 NAS-NRC (Washington, D.C., 1964).

2. Ibid., Foreword.

3. Ibid., p. 1.

4. Underseas mining, more efficient ocean shipping, weather forecasting, seaside recreation, sewage disposal.

5. A major emphasis is given to the process of discounting benefits and expenditures. Discounting is a device for taking into account the preference for current as against future income, hence reducing the value of future yields vis a vis current expenditures. Following one method of discounting the direct return on a 20-year investment in fisheries-oriented oceanographic research is calculated as more than four times larger and, by an alternative method, as nearly six times larger than if the same money had been invested at 10 percent compound interest. This is not the place to explain my fundamental objection to the assumptions underlying this procedure which applies a method taken in toto from the analysis of market transactions to the evaluation of public expenditures. As for discounting *method,* it can be left to actuaries and accountants; it is of no interest here. What is, for us, significant is the estimate of benefits, the assumptions implicit in those estimates, and the evidence on which they are based.

6. *Economic Benefits from Oceanographic Research,* pp. 13–16.

7. Ibid., pp. 13–16.

8. Ibid., pp. 15–16.

9. Though not germane to our central concern, it should be noted that the replacement of imports with domestic production is not a sufficient measure of net economic gain, as the authors of the report contend.

10. See Thomas Kuhn, *Structure of Scientific Revolutions* (Chicago: University of Chicago Press, 1962).

11. Gerhard Meseck, "Significance of the World Fishery for the Food Economy under Special Consideration of Asia", in *Proceedings of the International Seminar on Possibilities and Problems of Fisheries Development in Southeast Asia* (Berlin, 1969), p. 33.

12. W. Seeman. "German Fisheries, Their Post-War Development and Promotion by the Federation and the Federal States." *Proceedings,* p. 384.

13. Meseck, "Significance of the World Fishery . . . ," p. 33.

14. Report of the UN Secretary General of 24 April 1968. E/4487—"Marine Science and Technology: Survey and Proposals" (Part 1c), p. 66.

15. Ibid., p. 63.

16. Estimated by projecting potentials given in the Hsinhua Report of Minister of

Aquatic Products, 7th Session of the National Peoples Congress on 5 July 1957.

17. S. W. Ling, "Role and Possibilities of Brackishwater Fish and Shrimp Farming in Southeast Asia," *Proceedings,* p. 353.

18. Ibid., p. 354.

19. Report of UN Secretary General, p. 67.

20. *Proceedings,* p. 354.

21. Ibid., p. 355.

22. Meseck, "Significance of the World Fishery . . . ," p. 39.

23. "Role and Possibilities . . . ," p. 357.

24. Kenneth Boulding, "The Economics of the Coming Spaceship Earth," *Environmental Quality in a Growing Economy* (Baltimore: Johns Hopkins Press, 1966), p. 3.

25. Kenneth Boulding, "The Prospects of Economic Abundance," *The Control of the Environment* (Amsterdam: North Holland Press, 1967).

26. Marshall Goldman and Robert Shoop, "What is Pollution?" in Marshall Goldman (ed.), *Controlling Pollution* (Englewood Cliffs, N.J.: Prentice-Hall, 1967).

27. R. Steinberg. "The Role and Possibilities of Bilateral Assistance in the Further Development of Fisheries in Developing Countries," *Proceedings,* p. 138.

28. Ibid., p. 139.

29. A German scientist, who at the time headed a very successful fisheries-research center, wrote me

> I myself would like to underline what you have said. . . . I could give examples of countries with an undeveloped fishery that have started "oceanographical research work"—which means that they look for the temperature each day. But even with this simple work there is no connection with fishing problems, but they feel quite satisfied. . . .
>
> The same can be said about the fisheries institutes in different countries; they are very often nothing more than academic playgrounds for the younger members of the ruling society. There is no connection with the "low-classed" fishermen.

The observed tendency of researchers in those institutes to become isolated, to "lose touch," to stagnate, suggests that in order to offset the intellectually deadening effect of the social environment of traditional societies, more can be done to encompass indigenous research activities within an integral body of world science and of world technical development. The notion of "continuous education" for scientists and engineers is peculiarly relevant and important here, as are international programs for the continuous and systematic exchange of research personnel. The systematic "sponsorship" relationship between leading research organizations in the United States and Europe and independent "sister" research institutes in developing countries might be extended. Such a relationship, however, may have a distorting effect where the powerful sponsoring organization draws its small sister into the wake of research paths which, though intriguing to the scientists, are not germane to developmental opportunities or to the needs of developing economies. The relationship makes sense when the sponsoring organization is itself oriented toward practical problem-solving and, also, is prepared to develop the means of discovering and of remaining continuously cognizant of the needs and circumstances of development in which its "sister institute" operates.

It should not be supposed, however, that local research institutes can by themselves constitute an economic capacity to respond to the potentialities created through science. This will require many other linkages between information and innovation.

30. In 1968, for example, the Thai marine fisheries was cited (Tiews) as employing some 60 researchers, 8 research cutters and other vessels, with a main laboratory, 3 substations and further expansion in progress. In Japan, it was reported (Miyake) there were 60 fishery high schools, offering training in various specialized fields of fisheries, and more than a dozen fishery colleges and fishery departments in universities. Since 1964 Korea has been operating a Deep-Sea Fishing Training Center to produce qualified skippers and engineers, with similar centers started in Malaysia, Singapore, and India, and with a regional center (Southeast Asia Marine Fisheries Training Center) in Thailand and Singapore. Taiwan (Lui Yong-chio) reported six fisheries vocational schools, two fisheries-oriented colleges (the Provincial College of Oceanography and the China Marine College), and a Fisheries Biology Section in the National Taiwan University. The Taiwan Fisheries Bureau offers technical training to upgrade the skills of fisherman. Specialized research is carried on at the Fisheries Biological Research Institute and the Keelung Fisheries Institute covering the study of bottom fish resources, fishing techniques, fish processing, fish biology, mariculture, exploration for fishing grounds. Indonesia operates two junior fisheries schools, four vocational high schools for freshwater fisheries, a fisheries academy at Djakarta, and facilities for fisheries at four universities.

31. *Proceedings*, p. 28. ". . . the first successful project of this type, which was conducted in 1967 and which is known as the "Thai/Malaysian/German trawling survey off the coast of the Malay Peninsula." Here a bilateral agreement was reached between Thailand and Malaysia under ASEAN (Association of SE Asian Nations). For the survey of the bottom fish resources on the east coast of the Malay Peninsula, the Thai government supplied their survey boat *Pramong II*, the crew, and several biologists, while the Malaysian government paid the running costs and also assigned a few scientists."

Index

221